Math Mammoth
Grade 3 Answer Keys

for the complete curriculum
(Light Blue Series)

Includes answer keys to:

- Worktext part A
- Worktext part B
- Tests
- Cumulative Reviews

By Maria Miller

Copyright 2017 - 2019 Maria Miller
ISBN 978-1-042715-52-8

EDITION 6/2019

All rights reserved. No part of this book may be reproduced or transmitted in any form or by any means, electronic or mechanical, or by any information storage and retrieval system, without permission in writing from the author.

Copying permission: For having purchased this book, the copyright owner grants to the teacher-purchaser a limited permission to reproduce this material for use with his or her students. In other words, the teacher-purchaser MAY make copies of the pages, or an electronic copy of the PDF file, and provide them at no cost to the students he or she is actually teaching, but not to students of other teachers. This permission also extends to the spouse of the purchaser, for the purpose of providing copies for the children in the same family. Sharing the file with anyone else, whether via the Internet or other media, is strictly prohibited.

No permission is granted for resale of the material.

The copyright holder also grants permission to the purchaser to make electronic copies of the material for back-up purposes.

If you have other needs, such as licensing for a school or tutoring center, please contact the author at
https://www.MathMammoth.com/contact.php

Contents

Answer Key, Worktext part A 5

Answer Key, Worktext part B 51

Answer key, Chapter 1 Test 101
Answer key, Chapter 2 Test 102
Answer key, Chapter 3 Test 103
Answer key, Chapter 4 Test 104
Answer key, Chapter 5 Test 105
Answer key, Chapter 6 Test 106
Answer key, Chapter 7 Test 107
Answer key, Chapter 8 Test 108
Answer key, Chapter 9 Test 109
Answer key, Chapter 10 Test 110

Answer key, End-of-the-Year Test 111

Answer key, Cumulative Review Chapters 1-2 119
Answer key, Cumulative Review Chapters 1-3 120
Answer key, Cumulative Review Chapters 1-4 121
Answer key, Cumulative Review Chapters 1-5 122
Answer key, Cumulative Review Chapters 1-6 123
Answer key, Cumulative Review Chapters 1-7 124
Answer key, Cumulative Review Chapters 1-8 125
Answer key, Cumulative Review Chapters 1-9 126
Answer key, Cumulative Review Chapters 1-10 127

Math Mammoth Grade 3-A Answer Key

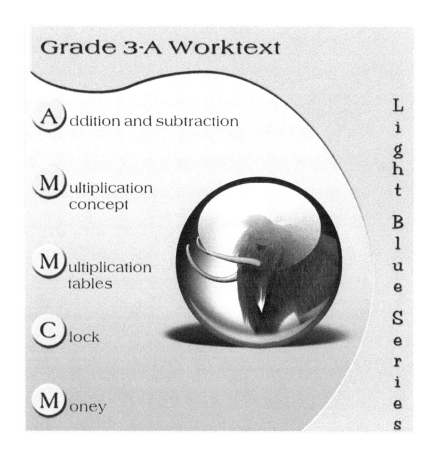

By Maria Miller

Copyright 2007-2019 Maria Miller.

Edition 6/2019

All rights reserved. No part of this book may be reproduced or transmitted in any form or by any means, electronic or mechanical, or by any information storage and retrieval system, without permission in writing from the author.

Copying permission: For having purchased this book, the copyright owner grants to the teacher-purchaser a limited permission to reproduce this material for use with his or her students. In other words, the teacher-purchaser MAY make copies of the pages, or an electronic copy of the PDF file, and provide them at no cost to the students he or she is actually teaching, but not to students of other teachers. This permission also extends to the spouse of the purchaser, for the purpose of providing copies for the children in the same family. Sharing the file with anyone else, whether via the Internet or other media, is strictly prohibited.

No permission is granted for resale of the material.

The copyright holder also grants permission to the purchaser to make electronic copies of the material for back-up purposes.

If you have other needs, such as licensing for a school or tutoring center, please contact the author at
https://www.MathMammoth.com/contact.php

Contents

Work-text page / Answer key page

Chapter 1: Addition and Subtraction

	Work-text page	Answer key page
Mental Addition	11	8
Review: Mental Subtraction	14	9
More Mental Subtraction	17	9
Ordinal Numbers and Roman Numerals	20	10
More Mental Addition	23	10
Mental Subtraction with Three-Digit Numbers	26	11
Regrouping in Addition	29	12
Review: Regrouping in Subtraction	33	13
Regrouping Twice in Subtraction	36	13
Regrouping Twice in Subtraction, Pt 2	40	15
Regrouping with Zero Tens	43	16
Regrouping with Zero Tens, Part 2	46	17
Rounding 2-Digit Numbers to the Nearest Ten	49	17
Rounding 3-Digit Numbers to the Nearest Ten	51	18
The Connection with Addition and Subtraction	54	19
Mileage Chart	58	20
Order of Operations	60	20
Graphs	62	20
Review Chapter 1	65	21

Chapter 2: Multiplication Concept

	Work-text page	Answer key page
Many Times the Same Group	70	22
Multiplication and Addition	71	22
Multiplying on a Number Line	74	23
Multiplication as an Array	77	23
Order of Operations 1	79	24
Understanding Word Problems, Part 1	80	24
Understanding Word Problems, Part 2	83	25
Multiplication in Two Ways	85	25
Order of Operations 2	89	27
Multiplying by Zero	91	27
Mixed Review Chapter 2	93	28
Review Chapter 2	95	29

Chapter 3: Multiplication Tables

	Work-text page	Answer key page
Multiplication Table of 2	102	30
Multiplication Table of 4	105	31
Multiplication Table of 10	107	32
Multiplication Table of 5	109	32
More Practice and Review (Tables of 2, 4, 5, and 10)	112	33
Multiplication Table of 3	115	34
Multiplication Table of 6	118	35
Multiplication Table of 11	120	36
Multiplication Table of 9	123	37
Multiplication Table of 7	127	39
Multiplication Table of 8	129	39
Multiplication Table of 12	132	40
Mixed Review Chapter 3	134	41
Review Chapter 3	136	41

Chapter 4: Telling Time

	Work-text page	Answer key page
Review: Reading the Clock	142	43
Half and Quarter Hours	144	43
Review: Till and Past	146	44
How Many Minutes Pass	148	44
More on Elapsed Time	150	45
Practice	152	45
Clock to the Minute	153	45
Elapsed Time in Minutes	156	46
Using the Calendar	158	46
Mixed Review Chapter 4	160	46
Review Chapter 4	162	46

Chapter 5: Money

	Work-text page	Answer key page
Using the Half-Dollar	165	47
Dollars	167	47
Making Change	170	47
Mental Math and Money Problems	174	47
Solving Money Problems	177	48
Mixed Review Chapter 5	181	48
Review Chapter 5	183	48

Chapter 1: Addition and Subtraction

Mental Addition, p. 11

1. a. 13, 14, 17   b. 15, 25, 65   c. 18, 38, 78   d. 50, 53, 51

2.

a. 50 + 14 = *50 + 10 + 4* = *64*	b. 80 + 11 = 80 + 10 + 1 = 91	c. 50 + 39 = 50 + 30 + 9 = 89
d. 35 + 60 = 30 + 5 + 60 = 95	e. 10 + 5 + 21 = 10 + 5 + 20 + 1 = 36	f. 29 + 40 + 30 = 20 + 9 + 40 + 30 = 99

3. a. 58   b. 90   c. 91   d. 79   e. 72   f. 110

4.

a. Add 20.	b. Add 40.	c. Add 15.	d. Add 25.
20	40	15	25
40	80	30	50
60	120	45	75
80	160	60	100
100	200	75	125
120	240	90	150
140	280	105	175

5. a. 68, 69   b. 128, 127   c. 50, 52
   d. 236, 235   e. 76, 75   f. 96, 93
   g. 270, 277   h. 300, 302   i. 690, 688

6. a. 69   b. 90   c. 107   d. 32   e. 89   f. 110

7.

29 + ___ = 36	7
66 + ___ = 76	10
48 + ___ = 56	
50 + ___ = 56	9
87 + ___ = 96	6
70 + ___ = 76	8
68 + ___ = 76	

| 86 + ___ = 96 |
| 46 + ___ = 56 |
| 57 + ___ = 66 |
| 38 + ___ = 46 |
| 89 + ___ = 96 |
| 39 + ___ = 46 |
| 77 + ___ = 86 |

Puzzle corner.

a. △ = 7

Solution: △ + △ + 1 = 15.

△ + △ must equal 14.

Therefore, △ must equal 7.

b. △ = 3, □ = 8.

Solution:

□ + △ = 11

□ − △ = 5

Guess and check is a great strategy here. Take two numbers that add to 11. For example, 6 and 5. Then check their difference (subtract): 6 - 5 = 1, which does not match, since □ - △ should be 5. So... guess again.
Then, we can try next 8 and 3....
which ends up being the correct answer.

c. △ = 10, □ = 7

Solution:

□ + △ = 17

□ + □ = 14

If two squares are 14, then one square = 7.
Then we tackle the top equation.
7 + △ = 17. The triangle equals 10.

Review: Mental Subtraction, p. 14

1.

a.	b.	c.	d.	e.
12 − 5 = 7	13 − 8 = 5	14 − 5 = 9	15 − 6 = 9	16 − 7 = 9
12 − 7 = 5	13 − 4 = 9	14 − 7 = 7	15 − 8 = 7	16 − 9 = 7
12 − 8 = 4	13 − 5 = 8	14 − 9 = 5	15 − 9 = 6	16 − 8 = 8
12 − 6 = 6	13 − 6 = 7	14 − 6 = 8	15 − 7 = 8	
12 − 4 = 8	13 − 9 = 4	14 − 8 = 6		f.
12 − 9 = 3	13 − 7 = 6			17 − 8 = 9
12 − 3 = 9				17 − 9 = 8

2. a. 9, 49   b. 4, 84   c. 9, 29

3. a. 7, 27, 57   b. 4, 34, 74   c. 9, 49, 149   d. 8, 68, 668

4. a. 57   b. 64   c. 46   d. 68   e. 18   f. 38

5. a. 63   b. 21   c. 24   d. 50   e. 31   f. 31

6. a. 3, 3   b. 4, 4   c. 6, 6

7. 50 − 13 − 13 = 24 You have $24 left.

8. 50 − 13 − 13 − 13 = 11 You would have $11 left.

9. 15 − 7 + 10 = 18 There are 18 children playing on the playground now.

10. 400 + 200 + 200 = 800 ft. The lion chased the antelope for 800 feet.

Puzzle corner:

Half of 10 is 5. There are four letters in June so half of that would be 2.
Double 2 is 4. So the number would be 452.

September has 9 letters. October has 7 letters. November has 8 letters.
So, the three-digit number is 978.

More Mental Subtraction, p. 17

1.

a. 54 − 10 = 44	b. 567 − 20 = 547	c. 93 − 30 = 63
54 − 30 = 24	567 − 200 = 367	137 − 20 = 117
289 − 20 = 269	778 − 40 = 738	543 − 400 = 143
289 − 50 = 239	778 − 400 = 378	803 − 600 = 203

2.

a. 65 − 26 = 39 + 4 + 30 + 5	b. 83 − 35 = 48 + 5 + 40 + 3
26 30 60 65	35 40 80 83

c.	d.	e.	f.
56 − 28 = 28	72 − 18 = 54	54 − 37 = 17	74 − 55 = 19
55 − 24 = 31	82 − 46 = 36	91 − 57 = 34	63 − 34 = 29

More Mental Subtraction, cont.

3.

a. $22 + \underline{\ ?\ } = 30. Solution: Ben needs $8.	b. $\underline{\ ?\ } - 8 = 17$ Solution: Mom bought 25 bushes.
c. $5 + 5 + 20 = \underline{\ ?\ }$ Solution: Jill has $30 now.	d. $7 + \underline{\ ?\ } = 21$ Solution: The Burns drank 14 bottles of water.
e. $20 - \underline{\ ?\ } = 13$ Solution: The gift cost $7.	f. $\underline{\ ?\ } - 12 - 5 - 2 = 9$ Solution: There were 28 cookies originally.

4. a. 16, 13   b. 38, 17   c. 48, 43   d. 36, 45

5. a. 9   b. 67   c. 172

6.

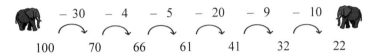

100   70   66   61   41   32   22

Puzzle corner:

a. △ = 22 □ = 8   b. △ = 21   c. □ = 10 △ = 3

Ordinal Numbers and Roman Numerals, p. 20

1.

2. a. 31st, thirty-first   b. 9th, ninth
   c. 12th, twelfth   d. 57th, fifty-seventh
   e. 99th, ninety-ninth   f. 52nd, fifty-second
   g. 61st, sixty-first   h. 43rd, forty-third

3. a. 2   b. 5   c. Jane is the fifth person in line.
   d. Mark is the second person in line.
   e. He is the eighth person in line.

4. a. 2, 7, 8, 12   b. 15, 21, 22, 35   c. 38, 53, 56, 61
   d. 63, 65, 80, 77   e. 83, 110, 107, 180

5. a. 4   b. 24   c. 29   d. 40
   e. 41   f. 49   g. 44   h. 93
   i. 74   j. 59   k. 85   l. 89
   m. 54   n. 56   o. 209   p. 94

6. a. XV; XVI; XVII   b. XXXI; XXXII; XXXIII
   c. XLII; XLIII; XLIV   d. L; LI; LII
   e. LXII; LXIII; LXIV   f. LXXV; LXXVI; LXXVII
   g. LXIX; LXX; LXXI   h. XCVII; XCVIII: XCIX

7. a. X   b. XX   c. LXXX   d. XXVIII
   e. XCIX   f. LXXXIX   g. XXX   h. L
   i. LXX   j. XCIV   k. XXVI   l. LIV

More Mental Addition, p. 23

1. a. 38, 138   b. 86, 686   c. 23, 323   d. 67, 267   e. 100, 500   f. 53, 753

2.

521	523	525	527	529
531	533	535	537	539
541	543	545	547	549
551	553	555	557	559

3. a. 129 + 9 = 138   The guitar cost $138 with tax.
   b. 150 − 138 = 12   Her change was $12.

4. a. 239, 484   b. 680, 532   c. 991, 300

More Mental Addition, cont.

5.

a.	b.	c.
95 + 5 = 100	293 + 6 = 299	994 + 5 = 999
95 + 6 = 101	293 + 7 = 300	995 + 5 = 1,000
95 + 7 = 102	293 + 8 = 301	996 + 5 = 1,001
95 + 8 = 103	293 + 9 = 302	997 + 5 = 1,002
95 + 9 = 104	293 + 10 = 303	998 + 5 = 1,003
95 + 10 = 105	293 + 11 = 304	999 + 5 = 1,004
95 + 11 = 106	293 + 12 = 305	1,000 + 5 = 1,005

6.

a.	b.	c.
393 + 8 = <u>401</u>	797 + 6 = 803	294 + 6 = 300
498 + 5 = 503	993 + 7 = 1,000	497 + 7 = 504
292 + 6 = 298	595 + 8 = 603	291 + 6 = 297

7. a. 129 + 7 = _?_   b. 132 − 5 = _?_ OR _?_ + 5 = 132

8. a. 18 + 22 = 40 They have 40 toy cars.
   b. Yes. They would each have 20 cars.

9. There are 30 days in the month of June, and July and August each have 31 days.
   Your vacation for those 3 months would be 30 + 31 + 31 = 92 days long.

10. a. 675, 500   b. 680, 888   c. 923, 201

11.

Number	4	8	7		2	6	2		5	3	9	9
Letter	Y	O	U		D	I	D		W	E	L	L

Mental Subtraction with Three-Digit Numbers, p. 26

1. a. 33, 133   b. 68, 268   c. 75, 675   d. 36, 636   e. 43, 343   f. 86, 686

2.

a. 152 − <u>6</u>
   152 − <u>2</u> − <u>4</u>
   = 146
	b. 244 − 9
	   244 − 4 − 5
	   = 235
		c. 823 − 8
		   823 − 3 − 5
		   = 815

d. 233 − 7 = 226	e. 191 − 5 = 186	f. 842 − 7 = 835

3. a. △ = 6   b. △ = 7   c. ▲ = 6

4.

a. 305 − 4 = 301	b. 340 − 8 = 332
305 − 5 = 300	340 − 10 = 330
305 − 6 = 299	340 − 12 = 328
305 − 7 = 298	340 − 14 = 326
305 − 8 = 297	340 − 16 = 324
305 − 9 = 296	340 − 18 = 322
305 − 10 = 295	340 − 20 = 320
305 − 11 = 294	340 − 22 = 318

Mental Subtraction with Three-Digit Numbers, cont.

5. a. 238 + 9 + 10 = 257   The new rent will be $257.
   b. 185 − 8 − 7 = 170   There were 170 people at the meeting.
   c. 125 + 6 = 131   Ernie traveled 131 kilometers.
   d. 510 − 11 = 499   The computer cost $499.

6. How can you get a into a refrigerator?

Number	0	7	1	6		10	5	1		4	0	0	8
Letter	O	P	E	N		T	H	E		D	O	O	R

,

Number	7	3	10		10	5	1			2	6
Letter	P	U	T		T	H	E			I	N

,

Number	9	5	3	10		10	5	1		4	0	0	8
Letter	S	H	U	T		T	H	E		D	O	O	R

Regrouping in Addition, p. 29

1. a. 231   b. 421   c. 532

2. a. 733   b. 642   c. 722   d. 845

3. a. 560   b. 911   c. 859   d. 748

4. a. 772   b. 533   c. 629

5.

	Fruits we Picked
= 10 oranges oranges	🍊🍊🍊
= 10 mangos mangos	🥭
= 10 bananas bananas	🍌🍌🍌🍌
= 10 apples apples	🍎🍎

6. a. 755 + 78 = 833   The second computer costs $833.
   b. 285 − 125 = 160   They still have to drive 160 kilometers.
   c. 35 + 19 + 22 = 76   Dad drove 76 kilometers.

7. a. 192 + 192 + 48 = 432   There are 432 puzzle pieces in the pile.
   b. 300 + 300 + 300 − 12 = 888   There are now 888 candles in total.
   c. 50 + 50 + 35 + 35 + 25 = 195   There are 195 kg of vegetables on the truck.

Puzzle corner:

```
   3 1 9           2 8 6
 + 1 9 1         + 6 4 5
 -------         -------
   5 1 0           9 3 1
```

Review: Regrouping in Subtraction, p. 33

1. a. 36   Check: 36 + 56 = 92   b. 17   Check: 17 + 38 = 55
   c. 606 Check: 606 + 156 = 762   d. 239 Check: 239 + 341 = 580

2. a. 272   b. 465   c. 581   d. 61

The number queue: 9 1 6 <u>5 8 1</u> 8 3 8 6 7 7 <u>6 1</u> 8 8 1 <u>2 7 2</u> 6 9 5 8 8 3 6 8 6 6 7 5 <u>4 6 5</u> 3 6 6

3. In the ones column, one take away one is not 9 - it is zero.   The correct way:
   He did not need to regroup (borrow) to be able   8 6 1
   to subtract one from one.   − 4 2 1
                                                                         ─────
                                                                         4 4 0

4. a. 172   Check: 172 + 357 = 529   b. 461   Check: 461 + 394 = 855
   c. 380   Check: 380 + 226 = 606   d. 96   Check: 96 + 541 = 637

5. a. 495 − 327 = 168   He still needs to save $168.
   b. 168 − 50 = 118   He still needs to save $118.
   c. Jayden has 15 + 3 = 18. Olivia has 21, so she has 3 more blocks than Jayden.
   d. 149 − 67 = 82   They still need to drive 82 kilometers.
   e. 178 + 210 = 388 They had 388 blue balls. 149 + 239 = 388 They had 388 red balls.

Regrouping Twice in Subtraction, p. 36

1.

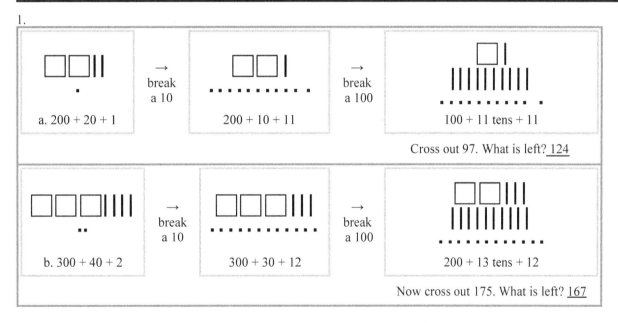

Regrouping Twice in Subtraction, cont.

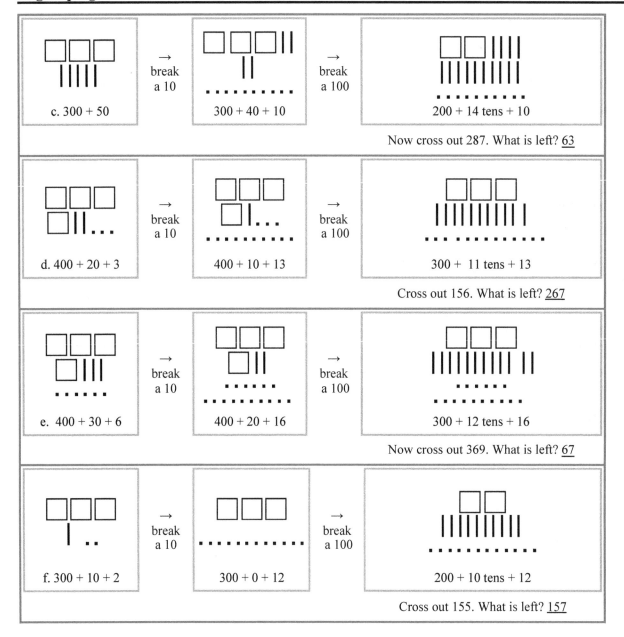

Regrouping Twice in Subtraction, cont.

2. a. 146   b. 268

a.	$600 + 20 + 3$ $- 400 - 70 - 7$	→	$600 + 10 + 13$ $- 400 - 70 - 7$	→	$500 + 110 + 13$ $- 400 - 70 - 7$ $100 + 40 + 6$
b.	$800 + 50 + 2$ $- 500 - 80 - 4$	→	$800 + 40 + 12$ $- 500 - 80 - 4$	→	$700 + 140 + 12$ $- 500 - 80 - 4$ $200 + 60 + 8$

3. a. 197   b. 68   c. 159

a.	$700 + 40 + 6$ $- 500 - 40 - 9$	→	$700 + 30 + 16$ $- 500 - 40 - 9$	→	$600 + 130 + 16$ $- 500 - 40 - 9$ $100 + 90 + 7$
b.	$400 + 60 + 1$ $- 300 - 90 - 3$	→	$400 + 50 + 11$ $- 300 - 90 - 3$	→	$300 + 150 + 11$ $- 300 - 90 - 3$ $ + 60 + 8$
c.	$900 + 10 + 4$ $- 700 - 50 - 5$	→	$900 + 0 + 14$ $- 700 - 50 - 5$	→	$800 + 100 + 14$ $- 700 - 50 - 5$ $100 + 50 + 9$

4. a. $32 + 32 + 32 = 96$ Yes, you can buy three backpacks and will have $4 left over.
   b. $57 - 12 + 45 = 90$ The total bill would be $90.

Regrouping Twice in Subtraction, Part 2, p. 40

1. a. 147   b. 469   c. 95   d. 165   e. 776   f. 197   g. 386   h. 188

2. a. $365 - 168 = 197$ There were 197 days without rain in 2010.
   b. $168 - 29 = 139$ There were 139 days that it did rain in 2010.
   c. $267 + 125 = 392$ Uptown School now has 392 students.
      $650 - 125 = 525$ Downtown School now has 525 students.
      $525 - 392 = 133$ Downtown School still has the most students. They have 133 more students than Uptown School.
   d. $775 - 250 - 180 = 345$ There are 345 striped shirts.

3. a. 655, $655 + 156 = 811$ b. 366, $366 + 277 = 643$
   c. 199, $199 + 266 = 465$ d. 254, $254 + 657 = 911$

4. a. 499   b. 556   c. 179   d. 367   e. 277   f. 258   g. 166   h. 379

Puzzle corner:

	4	3	8			8	5	3			6	1	9			6	8	4
−	1	2	3		−	3	3	6		−	3	5	5		−	4	7	7
	3	1	5			5	1	7			2	6	4			2	0	7

Regrouping with Zero Tens, p. 43

1. a. 172   b. 39   c. 166

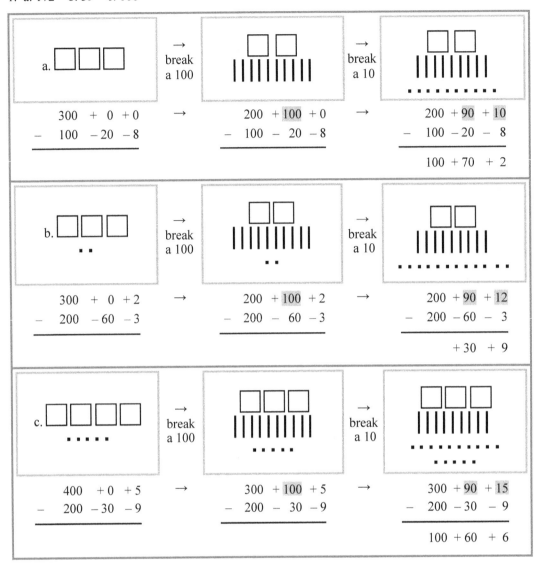

2. a. 176   b. 319   c. 184

3. Both 27 + _?_ = 83 *and* 83 − _?_ = 27 match the problem. Solution: there are 56 boys.

4. Answers will vary. Please check the student's work.

   22 + △ = 61. Answer: 39 are brown
   Example: There are 61 horses in the race.
   Some of the horses are brown and 22 are white.
   How many horses are brown?

Regrouping with Zero Tens, Part 2, p. 46

1. a. 437   b. 275   c. 78   d. 167   e. 789   f. 128   g. 398   h. 158

2. a. 245   b. 324   c. 259   d. 448

3. a. 665 + 125 = 790   Annie earned $790.
   b. 129 − 20 + 109 = 218   The two cameras will cost $218.
   c. 300 − 65 − 125 = 110   There are 110 red sweaters.

4. a. 97   b. 125   c. 115   d. 41   e. 60   f. 91

5. a. 25   b. 43   c. 29

6. a. 389   b. 269   c. 359   d. 265   e. 92   f. 726   g. 149   h. 158

Puzzle corner:

```
  6 0 8         8 0 0         6 0 1         6 1 0
− 2 9 3       − 2 3 6       − 3 5 7       − 4 0 3
-------       -------       -------       -------
  3 1 5         5 6 4         2 4 4         2 0 7
```

Rounding 2-Digit Numbers to the Nearest Ten, p. 49

1.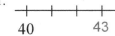

   40   43   46   48   50   52   57   59   60

   a. 52 is closest to 50   b. 57 is closest to 60   c. 43 is closest to 40
   d. 48 is closest to 50   e. 59 is closest to 60   f. 46 is closest to 50

   0   3   4   7   10   12   14   19   20

   g. 7 is closest to 10   h. 3 is closest to 0   i. 14 is closest to 10
   j. 19 is closest to 20   k. 12 is closest to 10   l. 4 is closest to 0

2. a. 30   b. 50   c. 60   d. 90   e. 10   f. 30   g. 70   h. 90

3. a. 40   b. 70   c. 100   d. 80   e. 10   f. 70   g. 80   h. 40

4.

a. a skirt, $28, and pants, $33
   a skirt about $30
   pants about $30
   together about $60
	b. a bicycle, $56, and light, $12
	   bicycle about $60
	   light about $10
	   together about $70
		c. a puzzle, $17, and book, $9
		   puzzle about $20
		   book about $10
		   together about $30

5. 50 + 20 ≈ 70 sacks of apples.

6. 10 + 20 + 20 ≈ $50 for the DVDs.

Rounding 3-Digit Numbers to the Nearest Ten, p. 51

1. a. 240 b. 290 c. 250 d. 300 e. 270 f. 210 g. 260 h. 300 i. 310 j. 300 k. 280 l. 240

2. a. 400, 900 b. 400, 390 c. 800, 810 d. 100, 1,000

3.

a. a computer, $296, and desk, $188 computer about $300 desk about $190 total bill about $490	b. a tennis racket, $123, and balls, $38 racket about $120 balls about $40 total bill about $160
c. total bill about $160	d. total bill about $200

4.

a. 470, 472, 480 472 ≈ 470	b. 820, 829, 830 829 ≈ 830	c. 510, 514, 520 514 ≈ 510
d. 310, 317, 320 317 ≈ 320	e. 600, 608, 610 608 ≈ 610	f. 450, 455, 460 455 ≈ 460

5. $130. 2 weeks = $60; 3 weeks = $90. It will take 5 weeks to earn enough to buy a bicycle.

6. 270 + 230 + 120 = 620 ft.   Estimate: 620 ft
   266 + 227 + 121 = 614 ft.

7. Across: Down:
   a. 633   a. 655
   b. 796   b. 819
   c. 447   c. 397
   d. 306   d. 512
             e. 911

a. 6	3	0		
6			c. 4	
0		b. 8	0	0
		2	0	
c. 4	5	0		e. 9
	1			1
	0		d. 3	1	0

The Connection with Addition and Subtraction, p. 54

1.

total 790		total 99	
670	120	65	34

a. 670 + 120 = 790
   790 − 670 = 120
   790 − 120 = 670

b. 65 + 34 = 99
   99 − 65 = 34
   99 − 34 = 65

total 390		total 400	
200	190	199	201

c. 200 + 190 = 390
   390 − 190 = 200
   390 − 200 = 190

d. 199 + 201 = 400
   400 − 199 = 201
   400 − 201 = 199

total 95		total 1000	
28	67	560	440

e. 28 + 67 = 95
   95 − 28 = 67
   95 − 67 = 28

f. 440 + 560 = 1,000
   1,000 − 440 = 560
   1,000 − 560 = 440

2.

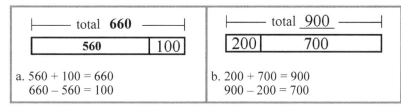

a. 560 + 100 = 660
   660 − 560 = 100

b. 200 + 700 = 900
   900 − 200 = 700

3.

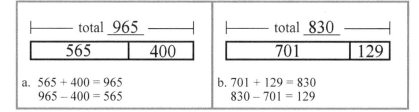

a. 565 + 400 = 965
   965 − 400 = 565

b. 701 + 129 = 830
   830 − 701 = 129

4.
  a. Ann needs 56 pins for a sewing project. She only has 41. How many more does she need?
     41 + 15 = 56
     56 − 41 = 15

  b. You are on page 224 of a book that has 380 pages. How many pages are left to read?
     224 + 156 = 380
     380 − 224 = 156

5. a. 39 b. 450 c. 23

6. a. 74 − 25 = 49   The difference in temperature is 49 degrees.
  b. 429 − 190 = 239   She needs $239 more.
  c. 80 + 42 = 122; 122 − 49 = 73   He needs to save $73 more.
  d. 30 + 30 + 30 = 90; 90 − 84 = 6   The cook needs to buy 3 cartons of eggs and will have six eggs left over.

7.

a. 199 + 35 = 234	b. 17 + 68 = 85
234 − 199 = 35	85 − 17 = 68

The Connection with Addition and Subtraction, cont.

8. a. 83 − 11 = 72; 72 − 45 = 27.   Jack has 27 more tennis balls than Robert.
   b. 37 + 15 = 52; 66 − 52 = 14.   He still needs $14.

9.

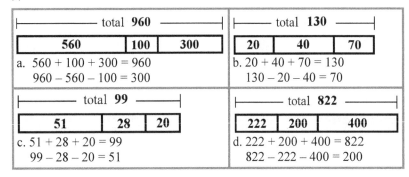

a. 560 + 100 + 300 = 960
   960 − 560 − 100 = 300

b. 20 + 40 + 70 = 130
   130 − 20 − 40 = 70

c. 51 + 28 + 20 = 99
   99 − 28 − 20 = 51

d. 222 + 200 + 400 = 822
   822 − 222 − 400 = 200

Mileage Chart, p. 58

1. 96 miles.
2. 189 miles.
3. 110 miles.
4. 284 + 284 = 568 miles.
5. 189 + 189 = 378 miles
6. 116 + 142 + 352 + 204 = 814 miles

7. a. He still had to drive 268 miles.
   b. He still had to drive 188 miles.
8. It is 55 miles further.
9. Almost, but not quite. In four hours he would drive
   45 + 45 + 45 + 45 = 180 miles of the 187-mile journey,
   so he would still have 7 miles farther to go.

Order of Operations, p. 60

1. a. 12, 16   b. 28, 28   c. 16, 12
2. a. 120 − (40 + 50)   b. 70 + 50 − 90 or (70 + 50) − 90
3. a. 0, 10   b. 140, 0
   c. 17, 25   d. 40, 80
4. a. 10 − (5 − 2) = 7   b. 20 − (5 − 2) − 1 = 16
   c. 15 − (5 + 2 − 1) = 9 d. 10 − (5 + 2) = 3
   e. 20 − (5 − 2 − 1) = 18 f. 15 − (5 + 2) − 1 = 7

5. a. 234 + 567 = 801 − 135 = 666
   b. 505 − 317 = 188 + 195 = 383
   c. 364 + (409 − 238) = 535
   d. 735 − (218 + 350) = 167

6. a. 380 + 380 − 25 = 735. The total cost was $735.
   b. 90 + 90 + 90 + 90 = 360; 360 − 125 = 235.
      He had $235 left.

Graphs, p. 62

1. a. Jane read the most books. She read 18 books.
   b. Jim read the fewest books. He read 8 books.
   c. Three more books. 16 − 13 = 3.
   d. The girls read a total of 18 + 15 + 9 + 12 = 54 books.
   e. The boys read a total of 14 + 8 + 16 + 13 = 51 books.
   f. The girls read more books; three more books.

2. a.

Vegetable use in one week	
Jacksons	🥕🥕🥕
Joneses	🥕
Millers	🥕🥕
Restaurant A	🥕🥕🥕🥕🥕
Restaurant B	🥕🥕🥕🥕🥕🥕

   b. Restaurant A used 35 kg of vegetables.
      Restaurant B used 40 kg of vegetables.
   c. 35 kg more   d. 75 kg total

Graphs, cont.

3. a.

Day	Mon	Tues	Wed	Thurs	Fri	Sat	Sun
Newspapers	about 30	about 40	about 40	about 40	about 50	about 30	about 70

   b. about 100 papers   c. about 40 more newspapers

4. a. 14 books   b. Annie; Lisa   c. 8 more books   d. 9 more books

e.

Books read	Jan	Feb	Mar	Apr	Total
Annie	13	21	18	14	66
Freddie	8	5	11	9	33
Lisa	8	13	16	18	55
Jonathan	10	8	14	15	47

   f. Annie read 33 more books than Freddie.

```
  Annie        Freddie        Lisa        Jonathan
   1 3            8            8            1 0
   2 1            5          1 3              8
   1 8          1 1          1 6            1 4
 + 1 4        +   9        + 1 8          + 1 5
   6 6          3 3          5 5            4 7
```

Review Chapter 1, p. 65

1. a. 308, 304   b. 230, 465   c. 994, 198

2. a. VI = 6	b. LVI = 56	c. LXV = 65	d. XLVIII = 48
e. 8 = VIII	f. 14 = XIV	g. 23 = XXIII	h. 67 = LXVII

3. a. 139   b. 294   c. 378   d. 377   e. 166

⊢ total 320 ⊣	⊢ total 900 ⊣
80 240	490 410
4. a. 80 + 240 = 320 320 − 80 = 240	b. 410 + 490 = 900 900 − 410 = 490

5.

a. 71 − 26 = 45 +4 +40 +1 26 → 30 → 70 → 71	b. 63 − 27 = 36 c. 82 − 51 = 31 d. 91 − 86 = 5

6. a. 31   b. 41   c. 510   d. 390

7. a. 12 ≈ 10   b. 677 ≈ 680   c. 46 ≈ 50

8. a. 27   b. 785

9. a. 800 − 270 − 270 = 260 yellow beads   b. 100 + 100 + 100 + (100 − 14) = 386 CDs

Chapter 2: Multiplication Concept

Many Times the Same Group, p. 70

1. b. 3 × 6   c. 4 × 0   d. 3 × 1   e. 5 × 2   f. 5 × 4

2. a.    b.   c.   d.   e.   f.

Multiplication and Addition, p. 71

1.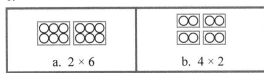

    a. 2 × 6   b. 4 × 2

2. a. 4 groups, 3 scissors in each.   4 × 3 scissors = 12 scissors;   3 + 3 + 3 + 3.
   b. 3 groups, 2 rams in each.   3 × 2 rams = 6 rams;   2 + 2 + 2.
   c. 3 groups, 1 dog in each.   3 × 1 dog = 3 dogs;   1 + 1 + 1.
   d. 1 group, 4 carrots in each.   1 × 4 carrots = 4 carrots.

3. a. 2 + 2 + 2 + 2 = 8. 4 × 2 = 8.   b. 1 + 1 + 1 + 1 + 1 = 5. 5 × 1 = 5.
   c. 2 + 2 + 2 + 2 + 2 = 10. 5 × 2 = 10.   d. 4 + 4 + 4 + 4 = 16. 4 × 4 = 16.

4.

a. Draw 3 groups of seven sticks.	b. Draw 2 groups of eight balls.
3 × 7 = 21	2 × 8 = 16
c. Draw 4 groups of four balls.	d. Draw 5 groups of two balls.
4 × 4 = 16	5 × 2 = 10

5.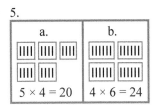

6. a. 5 × 4 = 20   b. 6 × 2 = 12
   c. 3 × 5 = 15   d. 3 × 11 = 33

Multiplying on a Number Line, p. 74

1. a. 7 × 2 = 14   b. 4 × 4 = 16   c. 3 × 3 = 9   d. 7 × 1 = 7

2.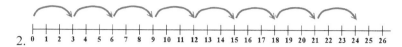

3. a. 15, 12   b. 24, 21   c. 18, 9   d. 6, 27

4. a. 8, 3   b. 6, 5   c. 7, 4   d. 2, 1

5.

6. a. 8, 16   b. 24, 28   c. 32, 12   d. 20, 4

7. a. 6, 2   b. 0, 3   c. 4, 2   d. 5, 1

8. a. 6 × 4 = 24
   b. 5 × 5 = 25
   c. 6 × 5 = 30
   d. 7 × 4 = 28
   e. 3 × 10 = 30

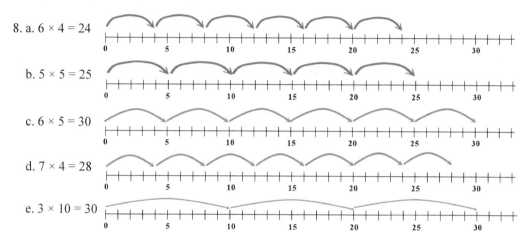

9. a. 6, 18, 20.   b. 10, 28, 24.   c. 30, 27, 40   d. 30, 22, 21.

Multiplication as an Array, p. 77

1. a. 2 rows, 5 carrots in each row.
     5 + 5
     2 × 5 = 10 carrots.
   b. 3 rows, 3 rams in each row.
     3 + 3 + 3
     3 × 3 = 9 rams.
   c. 2 rows, 1 bear in each row.
     1 + 1
     2 × 1 = 2 bears.
   d. 3 rows, 5 bulbs in each row.
     5 + 5 + 5
     3 × 5 = 15 bulbs.

2. a. 4 + 4 = 8; 2 × 4 = 8.
   b. 1 + 1 + 1 + 1 + 1 = 5; 5 × 1 = 5.
   c. 2 + 2 + 2 + 2 = 8; 4 × 2 = 8.
   d. 2 + 2 + 2 + 2 + 2 + 2 = 12; 6 × 2 = 12.
   e. 3 + 3 + 3 + 3 = 12; 4 × 3 = 12.
   f. 3 + 3 + 3 + 3 + 3 = 15; 5 × 3 = 15.
   g. 6 + 6 + 6 + 6 = 24; 4 × 6 = 24.
   h. 10 + 10 + 10 = 30; 3 × 10 = 30.
   i. 20 + 20 + 20 = 60; 3 × 20 = 60.
   j. 12 + 12 + 12 + 12 = 48; 4 × 12 = 48.

Order of Operations 1, p. 79

1. a. 19 b. 11 c. 40

2.

a. 5 + (4 × 2) 5 + 8 = __13__	b. (3 × 2) + 2 6 + 2 = 8	c. 20 − (4 × 4) 20 − 16 = 4
d. 15 + (3 × 2) 15 + 6 = 21	e. (3 × 5) − (2 × 4) 15 − 8 = 7	f. (2 × 5) + (1 × 4) 10 + 4 = 14
g. 5 + (1 × 2) + 5 5 + 2 + 5 = 12	h. 30 − (2 × 2) − 10 30 − 4 − 10 = 16	i. (5 × 1) + (2 × 3) 5 + 6 = 11
j. 10 + (5 × 4) 10 + 20 = 30	k. (2 × 6) + (2 × 7) 12 + 14 = 26	l. 50 − (3 × 2) + 6 50 − 6 + 6 = 50

Puzzle corner:

2 × 4 + 1 = 9   5 + 5 × 4 = 25   5 × 2 + 5 + 5 = 20

Understanding Word Problems, Part 1, p. 80

1. a. 4 × 6 = 24 tennis balls.
   b. 5 × 2 = 10 towels.
   c. 3 × 5 = 15 workers.
   d. 4 × 4 = 16 slices
   e. 4 × 5 = 20 flowers.
   f. 6 × 10 = 60 crayons

2. a. 4 × 2 + 3 × 5 = 23
   b. 3 × 4 + 4 × 3 = 24
   c. 5 × 4 + 7 = 27
   d. 4 × 5 + 3 × 6 + 8 = 46
   e. 4 × 10 + 2 × 6 + 7 = 59

3. a. 3 × 10 + 20 = $50
   b. 12 + 5 × 5 = $37
   c. 3 × 2 + 5 × 4 = $26
   d. 4 × 3 + 3 × 2 = $18

4. a. 5 × 2 + 5 = $15
   b. 4 × 10 + 2 × 20 = $80
   c. 2 + 2 + 4 × 3 = $16

Understanding Word Problems, Part 2, p. 83

1. a. 3 × 6 = 18   There are six students in each group.
   b. 9 × 2 = 18   She bought nine notebooks.
   c. 5 × 4 = 20   He can buy four soccer balls.
   d. 5 × 3 = 15   The total cost was $15.

2. a. 4 + 4 + 4 + 3 = 15 or 3 × 4 + 3 = 15   The people got 15 pizzas of pizza.
   b. 5 + 5 + 5 + 5 + 5 = 25 or 5 × 5 = 25   She will need five boxes.
   c. 6 + 6 + 6 + 6 + 6 = 30 or 5 × 6 = 30   She needs to learn 30 words.
      2 + 2 + 2 + 2 + 2 = 10 or 5 × 2 = 10   Ten words are in bold.
   d. 2 + 5 + 7 = 14   He bought 14 pieces of fruit.
   e. 4 + 4 + 4 + 4 + 3 = 19 or 4 × 4 + 3 = 19   There are 19 students in the class.

3.

a. 4 × 5 = 20 0 × 4 = 0 10 × 3 = 30	b. 10 × 0 = 0 6 × 3 = 18 1 × 78 = 78
c. 25 × 1 = 25 2 × 4 = 8 2 × 7 = 14	d. 0 × 49 = 0 10 × 1 = 10 2 × 6 = 12

4.

×	0	1	2	3	4
0	0	0	0	0	0
1	0	1	2	3	4
2	0	2	4	6	8
3	0	3	6	9	12
4	0	4	8	12	16

Multiplication in Two Ways, p. 85

1.

Five rows; each row has two rams. 2 + 2 + 2 + 2 + 2 rams 5 × 2 = 10	Two columns; each column has five rams. 5 + 5 rams 2 × 5 = 10
One row; it has five giraffes. 5 giraffes 1 × 5 = 5	Five columns, each column has one giraffe. 1 + 1 + 1 + 1 + 1 giraffes. 5 × 1 = 5

2. a. 4 + 4 = 2 × 4 = 8.   AND   2 + 2 + 2 + 2 = 4 × 2 = 8.
   b. 2 + 2 + 2 = 3 × 2 = 6.   AND   3 + 3 = 2 × 3 = 6.
   c. 1 × 3 = 3.   AND   1 + 1 + 1 = 3 × 1 = 3.
   d. 3 + 3 + 3 = 3 × 3 = 9. These facts are the same both ways. When the number of things in each row is the same as the number of things in each column (when the array is square), the facts are the same both ways.

Multiplication in Two Ways, cont.

3.

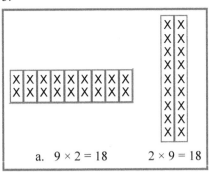   a. $9 \times 2 = 18$ $2 \times 9 = 18$

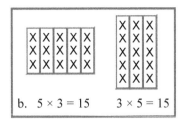   b. $5 \times 3 = 15$ $3 \times 5 = 15$

4. a. $5 \times 3 = 15$; $3 \times 5 = 15$.

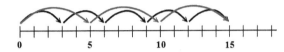

b. $7 \times 4 = 28$; $4 \times 7 = 28$.

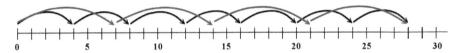

c. $6 \times 3 = 18$; $3 \times 6 = 18$.

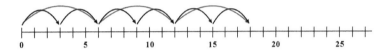

d. $7 \times 1 = 7$; $1 \times 7 = 7$.

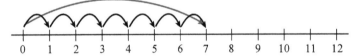

5.

a. $2 \times 10 = 20$	OR	$10 \times 2 = 20$	b. $7 \times 2 = 14$	OR	$2 \times 7 = 14$
Two groups of ten		Ten groups of two	Seven groups of two		Two groups of seven
c. $3 \times 4 = 12$	OR	$4 \times 3 = 12$	d. $11 \times 3 = 33$	OR	$3 \times 11 = 33$
Three groups of four		Four groups of three	Eleven groups of 3		Three groups of 11

6. The number line shows jumps of three.

$1 \times 3 = 3$	$4 \times 3 = 12$	$7 \times 3 = 21$	$10 \times 3 = 30$
$2 \times 3 = 6$	$5 \times 3 = 15$	$8 \times 3 = 24$	$11 \times 3 = 33$
$3 \times 3 = 9$	$6 \times 3 = 18$	$9 \times 3 = 27$	$12 \times 3 = 36$

7. a. $5 \times 4 = 20$ He used 20 rocks.
   b. $3 \times 12 = 36$ There are 36 pages in three booklets.
   c. $3 \times 4 = 12$ You can make three groups of sticks.
   d. $4 \times 5 = 20$ You can make four groups of sticks.

Order of Operations 2, p. 89

1. - 3. If there is anything in parentheses, do it first. Do the multiplications before additions or subtractions. Then, do the additions and subtractions from left to right. The first steps are highlighted.

1.

| a. 20 + 6 − 3 = 23 | c. 20 − 6 + 3 = 17 | e. 80 − 30 − (30 + 20) = 0 |
| b. 20 + (6 − 3) = 23 | d. 20 − (6 + 3) = 11 | f. 80 − (30 − 30) + 20 = 100 |

2.

a. 3 + 5 × 2 = 13	b. 5 × (3 + 1) = 20	c. 4 × (4 − 2) = 8
d. 3 × 6 − 11 = 7	e. 25 − 5 × 2 = 15	f. (3 − 2) × 6 = 6
g. (4 + 2) × 2 = 12	h. 3 × 5 + 2 × 4 = 23	
i. 2 × (4 + 3) + 8 = 22	j. 50 − (7 − 2) × 4 = 30	

3.

| a. 0 × 7 + 2 = 2 | b. 5 + 1 × 3 = 8 | c. 5 × (1 + 9) = 50 |
| d. (10 − 5) × 4 = 20 | e. 55 + 0 × 3 = 55 | f. 8 × 2 − 12 = 4 |

4.

| a. 3 × 4 − 2 × 3 = 6 | b. 6 + 7 × (4 − 2) = 20 |
| c. 2 × (5 + 4) + 5 = 23 | d. 30 − 2 − 7 × 2 = 14 |

5. a. 10 × 2 − 1 = 19   There are 19 plates on the table.
   b. 5 × 2 + 4 × 4 = 26   Twenty-six people can be seated in the restaurant.

Puzzle corner:

16 × 1 − 1 = 15   10 + 5 × 2 = 20   3 + 4 × 5 + 6 = 29
35 − 5 × 4 = 15   5 × 7 + 6 = 41   9 × 3 − 5 × 2 = 17

Multiplying by Zero, p. 91

1. a. 0, 0   b. 1, 9   c. 0, 10   d. 6, 0

2. Table of zero		Table of one	
1 × 0 = 0	7 × 0 = 0	1 × 1 = 1	7 × 1 = 7
2 × 0 = 0	8 × 0 = 0	2 × 1 = 2	8 × 1 = 8
3 × 0 = 0	9 × 0 = 0	3 × 1 = 3	9 × 1 = 9
4 × 0 = 0	10 × 0 = 0	4 × 1 = 4	10 × 1 = 10
5 × 0 = 0	11 × 0 = 0	5 × 1 = 5	11 × 1 = 11
6 × 0 = 0	12 × 0 = 0	6 × 1 = 6	12 × 1 = 12

3. a. 4 × 6 − 2 = 22   There were 22 good eggs.
   b. 3 × 20 = 60   Mary has 60 things in the jars.

Multiplying by Zero, cont.

4.

a.	b.
$35 \times 1 = 35$	$6 \times 5 = 30$
$1 \times 1 = 1$	$1 \times 0 = 0$
$10 \times 3 = 30$	$67 \times 1 = 67$
c.	d.
$1 \times 45 = 45$	$7 \times 2 = 14$
$0 \times 1 = 0$	$0 \times 0 = 0$
$0 \times 99 = 0$	$0 \times 10 = 0$

5.

×	0	1	2	3	4	5
0	0	0	0	0	0	0
1	0	1	2	3	4	5
2	0	2	4	6	8	10
3	0	3	6	9	12	15
4	0	4	8	12	16	20
5	0	5	10	15	20	25

Mixed Review Chapter 2, p. 93

1. a. 3, 7 b. 15, 23 c. 9, 61 d. 24, 175

2.

a. $93 + 6 = 99$	b. $47 + 29 = 76$	c. $15 + 18 = 33$
$893 + 6 = 899$	$607 + 9 = 616$	$624 + 8 = 632$

3.

a. $161 - \underline{6}$	b. $332 - 5$	c. $773 - 8$
$161 - \underline{1} - \underline{5} = 155$	$322 - 2 - 3 = 327$	$773 - 3 - 5 = 765$

4.

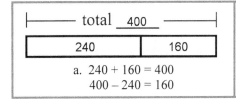
a. $240 + 160 = 400$
$400 - 240 = 160$

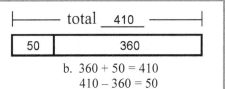
b. $360 + 50 = 410$
$410 - 360 = 50$

5.

a. $19 - (6 + 2) + 5 = 16$	b. $(800 - 60) - (50 - 40) = 730$
$19 - 6 + 2 + 5 = 20$	$800 - 60 - 50 - 40 = 650$

6. a. 259, 835   b. 176, 602

7. a. Danny ran 735 yards.
   b. They still have 174 km to go.
   c. The two jars have 580 beans.

8.

a. a toy, $28, and a set of books, $129	b. a ladder, $62, and wheelbarrow, $137
toy about $30	ladder about $60
set of books about $130	wheelbarrow about $140
together about $160	together about $200

Review Chapter 2, p. 95

1. a.    b.

2. a. $7 + 7 + 7 = 21$ b. $20 + 20 + 20 + 20 = 80$

3. a. $5 \times 4 = 20$
   b. $9 \times 3 = 27$

4.

a. $2 \times 2 = 4$ $1 \times 4 = 4$	b. $2 \times 10 = 20$ $3 \times 3 = 9$	c. $12 \times 0 = 0$ $12 \times 1 = 12$
d. $0 \times 5 = 0$ $2 \times 7 = 14$	e. $2 \times 40 = 80$ $3 \times 30 = 90$	f. $2 \times 400 = 800$ $1 \times 500 = 500$

5. a. 20 balls   b. $5 \times 4 = 20$ legs   c. $7 - 2 = 5 \times 2 = 10$ cans of cat food.   d. $7 - 1 = 6 \times 3 + 1 = 19$ books total

6. a. 10   b. 17   c. 23   d. 12

7.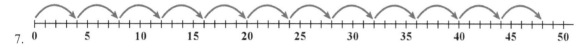

$1 \times 4 = 4$	$4 \times 4 = 16$	$7 \times 4 = 28$	$10 \times 4 = 40$
$2 \times 4 = 8$	$5 \times 4 = 20$	$8 \times 4 = 32$	$11 \times 4 = 44$
$3 \times 4 = 12$	$6 \times 4 = 24$	$9 \times 4 = 36$	$12 \times 4 = 48$

Chapter 3: Multiplication Tables

Multiplication Table of 2, p. 102

1. a. 0, 2, 4, 6, 8, 10, 12, 14, 16, 18, 20, 22, 24

2. a. 2, 4, 6, 8, 10, 12, 14, 16, 18, 20, 22, 24
   b. 1, 2, 3, 4, 5, 6, 7, 8, 9, 10, 11, 12

3.
6 × 2 = 12	7 × 2 = 14	2 × 3 = 6	2 × 7 = 14	2 × 8 = 16
9 × 2 = 18	2 × 2 = 4	2 × 11 = 22	2 × 4 = 8	3 × 2 = 6
4 × 2 = 8	8 × 2 = 16	2 × 9 = 18	2 × 6 = 12	2 × 5 = 10
2 × 1 = 2	12 × 2 = 24	2 × 12 = 24	8 × 2 = 16	10 × 2 = 20

4.
7 × 2 = 14	6 × 2 = 12	3 × 2 = 6	6 × 2 = 12	11 × 2 = 22
9 × 2 = 18	8 × 2 = 16	9 × 2 = 18	4 × 2 = 8	5 × 2 = 10
4 × 2 = 8	12 × 2 = 24	7 × 2 = 14	10 × 2 = 20	12 × 2 = 24
8 × 2 = 16	1 × 2 = 2	11 × 2 = 22	2 × 2 = 4	3 × 2 = 6

5. a. 24, 7, 8   b. 16, 10, 12   c. 18, 0, 2   d. 22, 20, 0

6.

a. Double 8 $\underline{8}$ + $\underline{8}$ = 16 $\underline{2}$ × $\underline{8}$ = 16	b. Double 13 13 + 13 = 26 2 × 13 = 26	c. Double 15 15 + 15 = 30 2 × 15 = 30
d. Double 25 25 + 25 = 50 2 × 25 = 50	e. Double 32 32 + 32 = 64 2 × 32 = 64	f. Double 45 45 + 45 = 90 2 × 45 = 90

7.

2 × 12 = 24 2 × 13 = 26 2 × 14 = 28	2 × 15 = 30 2 × 16 = 32 2 × 17 = 34	2 × 18 = 36 2 × 19 = 38 2 × 20 = 40	2 × 21 = 42 2 × 22 = 44 2 × 23 = 46

8.

a. 14 is even 2 × 7	b. 7 is odd 2 × ___	c. 18 is even 2 × 9
d. 21 is odd 2 × ___	e. 30 is even 2 × 15	f. 34 is even 2 × 17

9. a. 2 × 7 = 14
   b. 2 × 5 + 4 = 14
   c. 2 × 4 + 2 = 10
   d. 3 × 4 + 5 × 2 = 22

10. Answers will vary. Please check the student's work.

11. a. 2 × 7 − 3 = 11   Eleven birds stayed in the trees.
    b. 6 × 2 − 8 = 4   He has $4 left.

Puzzle corner:
2 × 8 + 11 = 27   The airplane cost $27.

Multiplication Table of 4, p. 105

1. 0, 4, 8, 12, 16, 20, 24, 28, 32, 36, 40, 44, 48

2. a.
1 × 4 = 4	7 × 4 = 28
2 × 4 = 8	8 × 4 = 32
3 × 4 = 12	9 × 4 = 36
4 × 4 = 16	10 × 4 = 40
5 × 4 = 20	11 × 4 = 44
6 × 4 = 24	12 × 4 = 48

b.
1 × 4 = 4	7 × 4 = 28
2 × 4 = 8	8 × 4 = 32
3 × 4 = 12	9 × 4 = 36
4 × 4 = 16	10 × 4 = 40
5 × 4 = 20	11 × 4 = 44
6 × 4 = 24	12 × 4 = 48

3.
6 × 4 = 24	7 × 4 = 28	4 × 3 = 12	4 × 7 = 28	3 × 4 = 12	4 × 8 = 32
9 × 4 = 36	8 × 4 = 32	4 × 11 = 44	4 × 6 = 24	4 × 5 = 20	2 × 4 = 8
4 × 4 = 16	12 × 4 = 48	4 × 9 = 36	4 × 12 = 48	10 × 4 = 40	4 × 1 = 4

4.
11 × 4 = 44	3 × 4 = 12	7 × 4 = 28	12 × 4 = 48	6 × 4 = 24
8 × 4 = 32	9 × 4 = 36	11 × 4 = 44	1 × 4 = 4	4 × 4 = 16
2 × 4 = 8	6 × 4 = 24	5 × 4 = 20	10 × 4 = 40	12 × 4 = 48

5. 0, 2, <u>4</u>, 6, <u>8</u>, 10, <u>12</u>, 14, <u>16</u>, 18, <u>20</u>, 22, <u>24</u>
   0, <u>4</u>, <u>8</u>, <u>12</u>, <u>16</u>, <u>20</u>, <u>24</u>, 28, 32, 36, 40, 44, 48

Numbers in both tables	Using 2	Using 4
0	0 × 2	<u>0</u> × 4
4	<u>2</u> × 2	<u>1</u> × 4
8	<u>4</u> × 2	<u>2</u> × 4
12	<u>6</u> × 2	<u>3</u> × 4

Numbers in both tables	Using 2	Using 4
16	<u>8</u> × 2	<u>4</u> × 4
20	<u>10</u> × 2	<u>5</u> × 4
24	<u>12</u> × 2	<u>6</u> × 4

6. a. 4 × 7 = 28   Seven goats have 28 legs.
   b. 3 × 4 + 7 × 2 = 26   They have a total of 26 legs.
   c. 15 × 1 = 15   You can buy 15 pairs of cheap socks for $15.
   d. 5 × 3 = 15   You can buy five pairs of expensive socks for $15.
   e. 3 × 1 + 2 × 3 = 9   She spent nine dollars on socks.

Multiplication Table of 10, p. 107

1. 0, 10, 20, 30, 40, 50, 60, 70, 80, 90, 100, 110, 120

2.
1 × 10 = 10	7 × 10 = 70
2 × 10 = 20	8 × 10 = 80
3 × 10 = 30	9 × 10 = 90
4 × 10 = 40	10 × 10 = 100
5 × 10 = 50	11 × 10 = 110
6 × 10 = 60	12 × 10 = 120

1 × 10 = 10	7 × 10 = 70
2 × 10 = 20	8 × 10 = 80
3 × 10 = 30	9 × 10 = 90
4 × 10 = 40	10 × 10 = 100
5 × 10 = 50	11 × 10 = 110
6 × 10 = 60	12 × 10 = 120

...both in the table of two and the table of ten?   10 × 2 = 2 × 10
...both in the table of four and the table of ten?   10 × 4 = 4 × 10

3.
5 × 10 = 50	6 × 10 = 60	10 × 8 = 80	10 × 7 = 70	2 × 5 = 10
12 × 10 = 120	9 × 10 = 90	10 × 4 = 40	10 × 10 = 100	10 × 3 = 30
7 × 10 = 70	11 × 10 = 110	10 × 12 = 120	10 × 11 = 110	10 × 6 = 60

4.
3 × 10 = 30	2 × 10 = 20	8 × 10 = 80	4 × 10 = 40	9 × 10 = 90
1 × 10 = 10	4 × 10 = 40	9 × 10 = 90	11 × 10 = 110	3 × 10 = 30
6 × 10 = 60	5 × 10 = 50	10 × 10 = 100	7 × 10 = 70	12 × 10 = 120

5. a.-b. It could be 2 cats and 7 chickens, or 3 cats and 5 chickens, or 4 cats and 3 chickens.
   (You cannot really have just 1 cat or 1 chicken since it speaks of them in the plural.)

6. a. 48 b. 16 c. 7 d. 0 e. 20 f. 44

7. Check the table the student filled in.

Multiplication Table of 5, p. 109

1. 0, 5, 10, 15, 20, 25, 30, 35, 40, 45, 50, 55, 60

2. a.
1 × 5 = 5	7 × 5 = 35
2 × 5 = 10	8 × 5 = 40
3 × 5 = 15	9 × 5 = 45
4 × 5 = 20	10 × 5 = 50
5 × 5 = 25	11 × 5 = 55
6 × 5 = 30	12 × 5 = 60

b.
1 × 5 = 5	7 × 5 = 35
2 × 5 = 10	8 × 5 = 40
3 × 5 = 15	9 × 5 = 45
4 × 5 = 20	10 × 5 = 50
5 × 5 = 25	11 × 5 = 55
6 × 5 = 30	12 × 5 = 60

...both in the table of five and table of two? 2 × 5 = 5 × 2

...both in the table of five and table of four? 4 × 5 = 5 × 4

...both in the table of five and table of ten? 10 × 5 = 5 × 10

3.
6 × 5 = 30	7 × 5 = 35	5 × 3 = 15	5 × 7 = 35	5 × 10 = 50
9 × 5 = 45	12 × 5 = 60	5 × 11 = 55	5 × 4 = 20	3 × 5 = 15
4 × 5 = 20	8 × 5 = 40	5 × 9 = 45	5 × 6 = 30	5 × 5 = 25

4.
7 × 5 = 35	4 × 5 = 20	11 × 5 = 55	8 × 5 = 40	11 × 5 = 55
1 × 5 = 5	9 × 5 = 45	5 × 5 = 25	10 × 5 = 50	6 × 5 = 30
12 × 5 = 60	2 × 5 = 10	7 × 5 = 35	12 × 5 = 60	3 × 5 = 15

Multiplication Table of 5, cont.

5. Table of 5: <u>0</u>, 5, <u>10</u>, 15, <u>20</u>, 25, <u>30</u>, 35, <u>40</u>, 45, <u>50</u>, 55, <u>60</u>
   Table of 10: <u>0</u>, <u>10</u>, <u>20</u>, <u>30</u>, <u>40</u>, <u>50</u>, <u>60</u>, 70, 80, 90, 100, 110, 120

Numbers in both tables	Using 5	Using 10
0	<u>0</u> × 5	<u>0</u> × 10
10	<u>2</u> × 5	<u>1</u> × 10
20	<u>4</u> × 5	<u>2</u> × 10
30	<u>6</u> × 5	<u>3</u> × 10

Numbers in both tables	Using 5	Using 10
40	<u>8</u> × 5	<u>4</u> × 10
50	<u>10</u> × 5	<u>5</u> × 10
60	<u>12</u> × 5	<u>6</u> × 10

6.
   a.
   10 × 2 + 0 = 20
   10 × 3 + 1 = 31
   10 × 4 + 2 = 42
   10 × 5 + 3 = 53
   10 × 6 + 4 = 64
   10 × 7 + 5 = 75
   10 × 8 + 6 = 86
   10 × 9 + 7 = 97
   10 × 10 + 8 = 108
   10 × 11 + 9 = 119

   b.
   5 × 1 + 1 = 6
   5 × 2 + 2 = 12
   5 × 3 + 3 = 18
   5 × 4 + 4 = 24
   5 × 5 + 5 = 30
   5 × 6 + 6 = 36
   5 × 7 + 7 = 42
   5 × 8 + 8 = 48
   5 × 9 + 9 = 54
   5 × 10 + 10 = 60

   c. The skip-counting pattern by 6.

7. Please check the students' work.

Puzzle corner

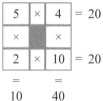

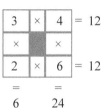

More Practice and Review (Tables of 2, 4, 5, and 10), p. 112

1. a. 18, 28, 20   b. 10, 12, 8   c. 14, 36, 40   d. 16, 48, 6
   e. 24, 16, 22   f. 8, 24, 2   g. 12, 44, 10   h. 32, 4, 4

2. a. 3 × 2 = 6 scoops.
   b. 2 × 12 − 4 = 20 eggs.
   c. 11 × 3 + 2 × 9 = 33 + 18 = 51 workers.
   d. 4 × 5 + 3 = 23 dolphin figurines

3. a. 45, 35, 50   b. 80, 40, 100   c. 30, 10, 55
   d. 70, 10, 110   e. 60, 120, 20   f. 40, 60, 5
   g. 50, 90, 30   h. 25, 20, 15

4. a. 4 × 5 = 20; She had five groups.
   b. 3 × 4 = 12; She wrote four invitations.
   c. 7 × 10 − 1 = 69; There are 69 passengers.
   d. 6 × 4 = 24; They took 24 sandwiches.
      2 × 5 = 10; They took 10 bottles of water.

More Practice and Review (Tables of 2, 4, 5, and 10), cont.

5. If there is anything in parentheses, do it first. Do the multiplications before additions or subtractions. Then, do the additions and subtractions from left to right. The highlighted parts are what you do first.

a. $3 + 7 \times 5 = 38$	b. $10 \times 6 - 10 \times 3 = 30$	c. $5 \times (5 - 4) = 5$
d. $(4 + 2) \times 5 = 30$	e. $5 \times 4 + 12 \times 4 = 68$	f. $0 + 7 \times 2 - 4 = 10$

6.

a.
1	3
2	6
3	9
4	12
5	15
6	18
7	21
8	24
9	27

b.
5	50
6	60
7	70
8	80
9	90
10	100
11	110
12	120
13	130

c.
1	20
2	40
3	60
4	80
5	100
6	120
7	140
8	160
9	180

Puzzle corner. a. □ = 3, △ = 5 (or vice versa) b. □ = 12, △ = 2. c. □ = 4, △ = 6 (or vice versa).

Multiplication Table of 3, p. 115

1. 0, 3, 6, 9, 12, 15, 18, 21, 24, 27, 30, 33, 36

2. a.
$1 \times 3 = 3$	$7 \times 3 = 21$
$2 \times 3 = 6$	$8 \times 3 = 24$
$3 \times 3 = 9$	$9 \times 3 = 27$
$4 \times 3 = 12$	$10 \times 3 = 30$
$5 \times 3 = 15$	$11 \times 3 = 33$
$6 \times 3 = 18$	$12 \times 3 = 36$

b.
$1 \times 3 = 3$	$7 \times 3 = 21$
$2 \times 3 = 6$	$8 \times 3 = 24$
$3 \times 3 = 9$	$9 \times 3 = 27$
$4 \times 3 = 12$	$10 \times 3 = 30$
$5 \times 3 = 15$	$11 \times 3 = 33$
$6 \times 3 = 18$	$12 \times 3 = 36$

3.

$6 \times 3 = 18$	$7 \times 3 = 21$	$3 \times 3 = 9$	$3 \times 7 = 21$	$3 \times 8 = 24$
$9 \times 3 = 27$	$2 \times 3 = 6$	$3 \times 11 = 33$	$3 \times 4 = 12$	$3 \times 3 = 9$
$4 \times 3 = 12$	$8 \times 3 = 24$	$3 \times 9 = 27$	$3 \times 6 = 18$	$3 \times 5 = 15$
$3 \times 1 = 3$	$12 \times 3 = 36$	$3 \times 12 = 36$	$8 \times 3 = 24$	$10 \times 3 = 30$

4.

$5 \times 3 = 15$	$4 \times 3 = 12$	$9 \times 3 = 27$	$12 \times 3 = 36$	$10 \times 3 = 30$
$11 \times 3 = 33$	$12 \times 3 = 36$	$11 \times 3 = 33$	$1 \times 3 = 3$	$2 \times 3 = 6$
$3 \times 3 = 9$	$8 \times 3 = 24$	$9 \times 3 = 27$	$6 \times 3 = 18$	$7 \times 3 = 21$

Multiplication Table of 3, cont.

5. a.
12 × 2 = 24
13 × 2 = 26
14 × 2 = 28
15 × 2 = 30
16 × 2 = 32
17 × 2 = 34
18 × 2 = 36
19 × 2 = 38
20 × 2 = 40
21 × 2 = 42

b.
1 × 2 − 1 = 1
2 × 2 − 2 = 2
3 × 2 − 3 = 3
4 × 2 − 4 = 4
5 × 2 − 5 = 5
6 × 2 − 6 = 6
7 × 2 − 7 = 7
8 × 2 − 8 = 8
9 × 2 − 9 = 9
10 × 2 − 10 = 10

6. a. 4 × 3 = 12 dollars, which is not enough, and 5 × 3 = 15 dollars, so he will need to work five days.
   b. 5 × 3 + 5 = 20; John now has $20. 20 − 14 = 6; He has six dollars left.
   c. 6 + 4 × 3 = 18; Yes, he can buy a book for $16.
   d. 11 × 3 + 1 = 34; Mom is 34 years old.
   e. 10 × 3 + 1 = 31; He would have to buy 10 bunches of three and one extra rose.
   f. Answers will vary.

7. Please check the students' work.

Multiplication Table of 6, p. 118

1. 0, 6, 12, 18, 24, 30, 36, 42, 48, 54, 60, 66, 72

2. a.
1 × 6 = 6	7 × 6 = 42
2 × 6 = 12	8 × 6 = 48
3 × 6 = 18	9 × 6 = 54
4 × 6 = 24	10 × 6 = 60
5 × 6 = 30	11 × 6 = 66
6 × 6 = 36	12 × 6 = 72

b.
1 × 6 = 6	7 × 6 = 42
2 × 6 = 12	8 × 6 = 48
3 × 6 = 18	9 × 6 = 54
4 × 6 = 24	10 × 6 = 60
5 × 6 = 30	11 × 6 = 66
6 × 6 = 36	12 × 6 = 72

3 × 8 = 24 × 2 = 8 × 6 = 48

3.
9 × 6 = 54	8 × 6 = 48	6 × 8 = 48	6 × 5 = 30	3 × 6 = 18
2 × 6 = 12	10 × 6 = 60	6 × 12 = 72	6 × 7 = 42	6 × 6 = 36
4 × 6 = 24	3 × 6 = 18	6 × 9 = 54	6 × 2 = 12	6 × 4 = 24
11 × 6 = 66	12 × 6 = 72	6 × 11 = 66	6 × 6 = 36	7 × 6 = 42

4.
12 × 6 = 72	3 × 6 = 18	9 × 6 = 54	7 × 6 = 42	9 × 6 = 54
1 × 6 = 6	8 × 6 = 48	4 × 6 = 24	6 × 6 = 36	5 × 6 = 30
10 × 6 = 60	2 × 6 = 12	7 × 6 = 42	11 × 6 = 66	12 × 6 = 72

5. **Table of 3:** 0, 3, <u>6</u>, 9, <u>12</u>, 15, <u>18</u>, 21, <u>24</u>, 27, <u>30</u>, 33, <u>36</u>
   Table of 6: 0, <u>6</u>, <u>12</u>, <u>18</u>, <u>24</u>, <u>30</u>, <u>36</u>, 42, 48, 54, 60, 66, 72

Numbers in both tables	Using 3	Using 6
0	0 × 3	<u>0</u> × 6
6	<u>2</u> × 3	<u>1</u> × 6
12	<u>4</u> × 3	<u>2</u> × 6
18	<u>6</u> × 3	<u>3</u> × 6

Numbers in both tables	Using 3	Using 6
24	<u>8</u> × 3	<u>4</u> × 6
30	<u>10</u> × 3	<u>5</u> × 6
36	<u>12</u> × 3	<u>6</u> × 6

6. Please check the student's work.

Multiplication Table of 11, p. 120

1. 0, 11, 22, 33, 44, 55, 66, 77, 88, 99, 110, 121, 132

2. a.
1 × 11 = 11	7 × 11 = 77
2 × 11 = 22	8 × 11 = 88
3 × 11 = 33	9 × 11 = 99
4 × 11 = 44	10 × 11 = 110
5 × 11 = 55	11 × 11 = 121
6 × 11 = 66	12 × 11 = 132

b.
1 × 11 = 11	7 × 11 = 77
2 × 11 = 22	8 × 11 = 88
3 × 11 = 33	9 × 11 = 99
4 × 11 = 44	10 × 11 = 110
5 × 11 = 55	11 × 11 = 121
6 × 11 = 66	12 × 11 = 132

3.

5 × 11 = 55	2 × 11 = 22	11 × 7 = 77	11 × 3 = 33	11 × 5 = 55
12 × 11 = 132	8 × 11 = 88	11 × 12 = 132	11 × 10 = 110	11 × 11 = 121
9 × 11 = 99	7 × 11 = 77	11 × 4 = 44	11 × 4 = 44	11 × 9 = 99
3 × 11 = 33	6 × 11 = 66	11 × 11 = 121	11 × 8 = 88	11 × 6 = 66

4.

8 × 11 = 88	7 × 11 = 77	5 × 11 = 55	6 × 11 = 66	1 × 11 = 11
12 × 11 = 132	11 × 11 = 121	3 × 11 = 33	2 × 11 = 22	4 × 11 = 44
10 × 11 = 110	9 × 11 = 99	12 × 11 = 132	11 × 11 = 121	10 × 11 = 110

5.

a.	b.
8 × 5 = 40	1 × 10 + 1 × 5 = 15
10 × 5 = 50	2 × 10 + 2 × 5 = 30
12 × 5 = 60	3 × 10 + 3 × 5 = 45
14 × 5 = 70	4 × 10 + 4 × 5 = 60
16 × 5 = 80	5 × 10 + 5 × 5 = 75
18 × 5 = 90	6 × 10 + 6 × 5 = 90
20 × 5 = 100	7 × 10 + 7 × 5 = 105
22 × 5 = 110	8 × 10 + 8 × 5 = 120
24 × 5 = 120	9 × 10 + 9 × 5 = 135
26 × 5 = 130	10 × 10 + 10 × 5 = 150
28 × 5 = 140	11 × 10 + 11 × 5 = 165

6. Answers may vary. Possible answers are listed.

a. 1 × 20 = 20 4 × 5 = 20 2 × 10 = 20	b. 1 × 18 = 18 3 × 6 = 18 2 × 9 = 18	c. 1 × 36 = 36 2 × 18 = 36 3 × 12 = 36 4 × 9 = 36 6 × 6 = 36
d. 1 × 30 = 30 2 × 15 = 30 5 × 6 = 30 3 × 10 = 30	e. 1 × 12 = 12 2 × 6 = 12 3 × 4 = 12	f. 1 × 24 = 24 3 × 8 = 24 2 × 12 = 24 4 × 6 = 24

7. Please check the student's answers.

Mystery Number:
   a. 28   b. 11 or 121   c. 25

Multiplication Table of 9, p. 123

1. 0, 9, 18, 27, 36, 45, 54, 63, 72, 81, 90, 99, 108

2. a.

$1 \times 9 = 9$	$7 \times 9 = 63$	b.	$1 \times 9 = 9$	$7 \times 9 = 63$
$2 \times 9 = 18$	$8 \times 9 = 72$		$2 \times 9 = 18$	$8 \times 9 = 72$
$3 \times 9 = 27$	$9 \times 9 = 81$		$3 \times 9 = 27$	$9 \times 9 = 81$
$4 \times 9 = 36$	$10 \times 9 = 90$		$4 \times 9 = 36$	$10 \times 9 = 90$
$5 \times 9 = 45$	$11 \times 9 = 99$		$5 \times 9 = 45$	$11 \times 9 = 99$
$6 \times 9 = 54$	$12 \times 9 = 108$		$6 \times 9 = 54$	$12 \times 9 = 108$

What same multiplication fact is both in...

... the table of nine and table of two? $2 \times 9 = 9 \times 2$
... the table of nine and table of five? $5 \times 9 = 9 \times 5$
... the table of nine and table of three? $3 \times 9 = 9 \times 3$
... the table of nine and table of ten? $10 \times 9 = 9 \times 10$
... the table of nine and table of four? $4 \times 9 = 9 \times 4$
... the table of nine and table of eleven? $11 \times 9 = 9 \times 11$

3.

$5 \times 9 = 45$	$8 \times 9 = 72$	$9 \times 10 = 90$	$9 \times 5 = 45$	$9 \times 8 = 72$	$11 \times 9 = 99$
$9 \times 9 = 81$	$10 \times 9 = 90$	$9 \times 3 = 27$	$9 \times 7 = 63$	$1 \times 9 = 9$	$9 \times 2 = 18$
$12 \times 9 = 108$	$6 \times 9 = 54$	$9 \times 1 = 9$	$9 \times 4 = 36$	$9 \times 6 = 54$	$9 \times 9 = 81$

4.

$2 \times 9 = 18$	$4 \times 9 = 36$	$8 \times 9 = 72$	$12 \times 9 = 108$	$9 \times 9 = 81$
$5 \times 9 = 45$	$1 \times 9 = 9$	$10 \times 9 = 90$	$11 \times 9 = 99$	$8 \times 9 = 72$
$3 \times 9 = 27$	$8 \times 9 = 72$	$9 \times 9 = 81$	$7 \times 9 = 63$	$6 \times 9 = 54$

5.

Multiply:	Add the digits:	Multiply:	Add the digits:
$1 \times 9 = 9$	$0 + 9 = 9$	$7 \times 9 = 63$	$6 + 3 = 9$
$2 \times 9 = 18$	$1 + 8 = 9$	$8 \times 9 = 72$	$7 + 2 = 9$
$3 \times 9 = 27$	$2 + 7 = 9$	$9 \times 9 = 81$	$8 + 1 = 9$
$4 \times 9 = 36$	$3 + 6 = 9$	$10 \times 9 = 90$	$9 + 0 = 9$
$5 \times 9 = 45$	$4 + 5 = 9$	$11 \times 9 = 99$	$9 + 9 = 18$; $1 + 8 = 9$
$6 \times 9 = 54$	$5 + 4 = 9$	$12 \times 9 = 108$	$1 + 0 + 8 = 9$

6. The yellow numbers count up from 0 to 9.
   The blue numbers count down from 9 to 0.

$1 \times 9 = 0\ 9$
$2 \times 9 = 1\ 8$
$3 \times 9 = 2\ 7$
$4 \times 9 = 3\ 6$
$5 \times 9 = 4\ 5$
$6 \times 9 = 5\ 4$
$7 \times 9 = 6\ 3$
$8 \times 9 = 7\ 2$
$9 \times 9 = 8\ 1$
$10 \times 9 = 9\ 0$

Multiplication Table of 9, cont.

7. The table of 3: **0**, 3, 6, **9**, 12, 15, **18**, 21, 24, **27**, 30, 33, **36**
   The table of 9: **0**, **9**, **18**, **27**, **36**, 45, 54, 63, 72, 81, 90, 99, 108

Numbers in both tables	Using 3	Using 9
0	0 × 3	<u>0</u> × 9
9	<u>3</u> × 3	<u>1</u> × 9
18	<u>6</u> × 3	<u>2</u> × 9

Numbers in both tables	Using 3	Using 9
27	<u>9</u> × 3	<u>3</u> × 9
36	<u>12</u> × 3	<u>4</u> × 9

8.

Table of 3:	0, 3, 6, 9, 12, 15, 18, 21, 24, 27, 30, 33, 36, 39, 42, 45, 48, 51, 54, 57, 60, 63, 66, 69, 72
Table of 9:	0, 9, 18, 27, 36, 45, 54, 63, 72

Every number in the table of <u>9</u> is also in the table of <u>3</u>.

9.

Multiply:	Add the digits:
10 × 9 = 90	9 + 0 = 9
11 × 9 = 99	9 + 9 = 18 ; 1 + 8 = 9
12 × 9 = 108	1 + 0 + 8 = 9
13 × 9 = 117	1 + 1 + 7 = 9
14 × 9 = 126	1 + 2 + 6 = 9
15 × 9 = 135	1 + 3 + 5 = 9

Multiply:	Add the digits:
16 × 9 = 144	1 + 4 + 4 = 9
17 × 9 = 153	1 + 5 + 3 = 9
18 × 9 = 162	1 + 6 + 2 = 9
19 × 9 = 171	1 + 7 + 1 = 9
20 × 9 = 180	1 + 8 + 0 = 9
21 × 9 = 189	1 + 8 + 9 = 18; 1 + 8 = 9

10. Please check the student's work.

Puzzle corner. The comparison you cannot do is marked with a ?. Since we do not know the value of △, △ × 1, which equals △, could be more, less, or equal to 70.

9 × △ < 10 × △	△ × 5 > △ × 4	△ × 0 < 3 × 6
△ × 8 > △ × 4	4 × △ < △ × 8	△ × 1 ? 10 × 7
△ × 8 > △ × 5	△ × 2 = △ + △	△ × 3 > △ + △

Multiplication Table of 7, p. 127

1. 0, 7, 14, 21, 28, 35, 42, 49, 56, 63, 70, 77, 84

2. a.
1 × 7 = 7	7 × 7 = 49
2 × 7 = 14	8 × 7 = 56
3 × 7 = 21	9 × 7 = 63
4 × 7 = 28	10 × 7 = 70
5 × 7 = 35	11 × 7 = 77
6 × 7 = 42	12 × 7 = 84

b.
1 × 7 = 7	7 × 7 = 49
2 × 7 = 14	8 × 7 = 56
3 × 7 = 21	9 × 7 = 63
4 × 7 = 28	10 × 7 = 70
5 × 7 = 35	11 × 7 = 77
6 × 7 = 42	12 × 7 = 84

3.

9 × 7 = 63	8 × 7 = 56	7 × 8 = 56	7 × 5 = 35	3 × 7 = 21
4 × 7 = 28	10 × 7 = 70	7 × 12 = 84	7 × 7 = 49	6 × 7 = 42
11 × 7 = 77	7 × 6 = 42	7 × 9 = 63	7 × 2 = 14	4 × 7 = 28
5 × 7 = 35	10 × 7 = 70	6 × 7 = 42	4 × 7 = 28	8 × 7 = 56
11 × 7 = 77	3 × 7 = 21	8 × 7 = 56	12 × 7 = 84	7 × 7 = 49
6 × 7 = 42	2 × 7 = 14	5 × 7 = 35	5 × 7 = 35	9 × 7 = 63

4. a. 4 × 7 = 28; Jenny used four boxes.
   b. 12 × 2 = 24 socks.
   c. 3 × 12 − 8 = 28 eggs.
   d. 5 × 6 = 30; You need five tables to seat your dinner guests.

5. Please check the students' work.

Multiplication Table of 8, p. 129

1. 0, 8, 16, 24, 32, 40, 48, 56, 64, 72, 80, 88, 96

2. a.
1 × 8 = 8	7 × 8 = 56
2 × 8 = 16	8 × 8 = 64
3 × 8 = 24	9 × 8 = 72
4 × 8 = 32	10 × 8 = 80
5 × 8 = 40	11 × 8 = 88
6 × 8 = 48	12 × 8 = 96

b.
1 × 8 = 8	7 × 8 = 56
2 × 8 = 16	8 × 8 = 64
3 × 8 = 24	9 × 8 = 72
4 × 8 = 32	10 × 8 = 80
5 × 8 = 40	11 × 8 = 88
6 × 8 = 48	12 × 8 = 96

3.

8 × 8 = 64	9 × 8 = 72	8 × 4 = 32	8 × 5 = 40	8 × 8 = 64
8 × 6 = 48	8 × 11 = 88	8 × 12 = 96	7 × 8 = 56	8 × 10 = 80
3 × 8 = 24	8 × 6 = 48	2 × 8 = 16	8 × 9 = 72	8 × 6 = 48

4.

4 × 8 = 32	3 × 8 = 24	11 × 8 = 88	5 × 8 = 40	8 × 8 = 64
1 × 8 = 8	6 × 8 = 48	9 × 8 = 72	7 × 8 = 56	12 × 8 = 96
8 × 8 = 64	2 × 8 = 16	10 × 8 = 80	6 × 8 = 48	11 × 8 = 88

5. Table of 4: **0**, 4, **8**, 12, **16**, 20, **24**, 28, **32**, 36, **40**, 44, **48**
   Table of 8: **0**, **8**, **16**, **24**, **32**, **40**, **48**, 56, 64, 72, 80, 88, 96

Numbers in both tables	Using 4	Using 8
0	0 × 4	0 × 8
8	2 × 4	1 × 8
16	4 × 4	2 × 8
24	6 × 4	3 × 8

Numbers in both tables	Using 4	Using 8
32	8 × 4	4 × 8
40	10 × 4	5 × 8
48	12 × 4	6 × 8

Multiplication Table of 8, cont.

6. Table of 4: 0, 4, 8, 12, 16, 20, 24, 28, 32, 36, 40, 44, 48, 52, 56, 60, 64, 68, 72, 76, 80, 84, 88, 92, 96.
   Table of 8: 0, 8, 16, 24, 32, 40, 48, 56, 64, 72, 80, 88, 96
   <u>Every second number in the table of 4 is found in the table of 8.</u>

7. a. 8 × 5 = 40. There are 40 erasers in five packages.
   b. 8 × 3 = 24. She needs three packages of erasers so each child can have one.
   c. 2 × 5 = 10. It will take them five weeks to eat ten kilograms of beans.

8. Please check the student's work.

Puzzle corner. a. □ = 6, △ = 8 (or vice versa) b. □ = 4, △ = 12 (or vice versa). c. □ = 12, △ = 3.

Multiplication Table of 12, p. 132

1. 0, 12, 24, 36, 48, 60, 72, 84, 96, 108, 120, 132, 144

2. a.
1 × 12 = 12	7 × 12 = 84
2 × 12 = 24	8 × 12 = 96
3 × 12 = 36	9 × 12 = 108
4 × 12 = 48	10 × 12 = 120
5 × 12 = 60	11 × 12 = 132
6 × 12 = 72	12 × 12 = 144

   b.
1 × 12 = 12	7 × 12 = 84
2 × 12 = 24	8 × 12 = 96
3 × 12 = 36	9 × 12 = 108
4 × 12 = 48	10 × 12 = 120
5 × 12 = 60	11 × 12 = 132
6 × 12 = 72	12 × 12 = 144

3.
3 × 12 = 36	9 × 12 = 108	12 × 4 = 48	12 × 1 = 12	7 × 12 = 84
2 × 12 = 24	10 × 12 = 120	12 × 5 = 60	12 × 7 = 84	12 × 3 = 36
1 × 12 = 12	6 × 12 = 72	12 × 8 = 96	12 × 9 = 108	4 × 12 = 48
8 × 12 = 96	12 × 12 = 144	12 × 11 = 132	12 × 6 = 72	12 × 2 = 24

4.
3 × 12 = 36	2 × 12 = 24	7 × 12 = 84	6 × 12 = 72	12 × 12 = 144
1 × 12 = 12	4 × 12 = 48	12 × 12 = 144	10 × 12 = 120	11 × 12 = 132
6 × 12 = 72	5 × 12 = 60	8 × 12 = 96	5 × 12 = 60	9 × 12 = 108

5.

×	0	1	2	3	4	5	6	7	8	9	10	11	12
0	0	0	0	0	0	0	0	0	0	0	0	0	0
1	0	1	2	3	4	5	6	7	8	9	10	11	12
2	0	2	4	6	8	10	12	14	16	18	20	22	24
3	0	3	6	9	12	15	18	21	24	27	30	33	36
4	0	4	8	12	16	20	24	28	32	36	40	44	48
5	0	5	10	15	20	25	30	35	40	45	50	55	60
6	0	6	12	18	24	30	36	42	48	54	60	66	72
7	0	7	14	21	28	35	42	49	56	63	70	77	84
8	0	8	16	24	32	40	48	56	64	72	80	88	96
9	0	9	18	27	36	45	54	63	72	81	90	99	108
10	0	10	20	30	40	50	60	70	80	90	100	110	120
11	0	11	22	33	44	55	66	77	88	99	110	121	132
12	0	12	24	36	48	60	72	84	96	108	120	132	144

Mixed Review Chapter 3, p. 134

1. a. 27   b. 687   c. 5

2. a. 660   b. 600   c. 820   d. 60

3. a. 746   b. 721

4. 16 + 36 = 52   or 52 − 16 = 36   There are 36 white candles.

5.

| a. 12 is XII | b. 34 is XXXIV | c. 55 is LV | d. 80 is LXXX |

6.
a. △ = 27   b. △ = 700   c. ▲ = 430

7.

| a. 3 × 5 | b. 4 × 3 |

8. a. 5 × 10 = 50   Five children have 50 toes all totaled.
   b. 3 × 5 = 15   He has three rows of cars.
   c. 5 × 4 = 20   You need five tables to seat 20 people.

9. a. The Sports Club is the most popular.
   b. There are 15 more students.
   c. There are 68 students in the three clubs.

Review Chapter 3, p. 136

1.

×	0	1	2	3	4	5	6	7	8	9	10	11	12
0	0	0	0	0	0	0	0	0	0	0	0	0	0
1	0	1	2	3	4	5	6	7	8	9	10	11	12
2	0	2	4	6	8	10	12	14	16	18	20	22	24
3	0	3	6	9	12	15	18	21	24	27	30	33	36
4	0	4	8	12	16	20	24	28	32	36	40	44	48
5	0	5	10	15	20	25	30	35	40	45	50	55	60
6	0	6	12	18	24	30	36	42	48	54	60	66	72
7	0	7	14	21	28	35	42	49	56	63	70	77	84
8	0	8	16	24	32	40	48	56	64	72	80	88	96
9	0	9	18	27	36	45	54	63	72	81	90	99	108
10	0	10	20	30	40	50	60	70	80	90	100	110	120
11	0	11	22	33	44	55	66	77	88	99	110	121	132
12	0	12	24	36	48	60	72	84	96	108	120	132	144

Review Chapter 3, cont.

2. a. 9×8 < 10×8 b. 9×5 > 11×4 c. 9×2 = 3×6

   d. 9×8 > 9×4 e. 4×4 = 2×8 f. 10×11 > 10×7

   g. 10×8 > 10×5 h. 9×2 < 4×5 i. 9×8 > 9×6

3.

$1 \times 3 = 3$	$7 \times 3 = 21$	$1 \times 6 = 6$	$7 \times 6 = 42$
$2 \times 3 = 6$	$8 \times 3 = 24$	$2 \times 6 = 12$	$8 \times 6 = 48$
$3 \times 3 = 9$	$9 \times 3 = 27$	$3 \times 6 = 18$	$9 \times 6 = 54$
$4 \times 3 = 12$	$10 \times 3 = 30$	$4 \times 6 = 24$	$10 \times 6 = 60$
$5 \times 3 = 15$	$11 \times 3 = 33$	$5 \times 6 = 30$	$11 \times 6 = 66$
$6 \times 3 = 18$	$12 \times 3 = 36$	$6 \times 6 = 36$	$12 \times 6 = 72$

Every other answer from the table of three is in the table of six.

4. a. $11 \times 7 = 77$ The girls have a total of 77 schoolbooks.
   b. $4 \times 5 = 20$ There will be five groups.
   c. $4 \times 3 + 7 = 19$ The total cost was $19.
   d. $12 \times 2 = 24$ He bought 12 packages of seed.
   e. $5 \times 4 + 3 \times 4 + 20 \times 2 = 72$ They have a total of 72 feet.

5. a. 3, 8, 5   b. 3, 11, 2   c. 7, 8, 9   d. 5, 9, 7
   e. 4, 7, 9   f. 12, 7, 9   g. 6, 4, 9   h. 5, 7, 9

Mystery numbers: a. 44.   b. 24   c. 29   d. 24   e. 44   f. 12

Chapter 4: Clock

Review: Reading the Clock, p. 142

1.

a.	b.	c.	d.
8 : 15	11 : 30	2 : 40	5 : 55
8 : 20	11 : 35	2 : 45	6 : 00
8 : 25	11 : 40	2 : 50	6 : 05
8 : 30	11 : 45	2 : 55	6 : 10
8 : 35	11 : 50	3 : 00	6 : 15
8 : 40	11 : 55	3 : 05	6 : 20
8 : 45	12 : 00	3 : 10	6 : 25
8 : 50	12 : 05	3 : 15	6 : 30
8 : 55	12 : 10	3 : 20	6 : 35
9 : 00	12 : 15	3 : 25	6 : 40

2. a. 1:40 b. 12:05 c. 8:25 d. 3:45 e. 7:35 f. 11:50 g. 7:20 h. 1:15

3.

	a. 3:35	b. 5:20	c. 4:50	d. 7:45
10 min. later →	3:45	5:30	5:00	7:55

4.

	a. 3:55	b. 4:05	c. 4:00	d. 7:00
5 min. earlier →	3:50	4:00	3:55	6:55

Half and Quarter Hours p. 144

1. a. a quarter past 2   b. a quarter till 6   c. a quarter till 12   d. half past 12
   e. a quarter past 1   f. a quarter till 5   g. 4 o'clock   h. a quarter past 9

2. a. 4:45   b. 12:15   c. 8:45   d. 3:15   e. 11:30   f. 11:45   g. 8:15   h. 12:45

3. a. half past 7   b. a quarter past 5   c. a quarter till 6   d. a quarter till 10
   e. a quarter past 12   f. half past 12   g. a quarter till 12   h. a quarter till 8

4. a. 4 o'clock   b. quarter till 8   c. quarter past 6   d. quarter past 5   e. half past 5   f. quarter till 7

5. a. 5:45 or a quarter till 6   b. 2:15 or a quarter past 2   c. 1:45 or a quarter till 2

Review: Till and Past, p. 146

1. a. a quarter till 6 b. ten past 11 c. a quarter past 5 d. a quarter till 12 e. 25 till 7
   f. 20 past 3 g. 20 till 9 h. a quarter till 1 i. 5 till 10

2. a. a quarter till 12 b. 25 past 7 c. 20 till 7 d. a quarter past 10 e. 5 past 8 f. 25 till 6

3. b. 6:50 c. 3:25 d. 5:30 e. 4:45 f. 5:35 g. 11:55
   h. 12:45 i. 11:45 j. 6:15 k. 10:55 l. 8:40

4. a. a quarter till 6 b. 5 till 12 c. a quarter past 2 d. a quarter till 3
   e. five till 4 f. 25 till 7 g. 25 past 8 h. a quarter past 10

How Many Minutes Pass, p. 148

1. a. 15 minutes b. 15 minutes c. 20 minutes d. 35 minutes

2. a. 10 minutes b. 15 minutes c. 5 minutes d. 10 minutes e. 25 minutes f. 20 minutes

3.

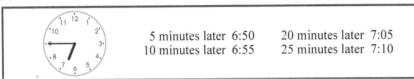

5 minutes later 6:50   20 minutes later 7:05
10 minutes later 6:55   25 minutes later 7:10

4. a. 45 minutes b. 35 minutes c. 20 minutes d. 15 minutes

5.

a. 5 minutes later 8:25   20 minutes later 8:40
   10 minutes later 8:30 35 minutes later 8:55

b. 10 minutes later 4:35 35 minutes later 5:00
   30 minutes later 4:55 45 minutes later 5:10

c. 10 minutes later 10:55 30 minutes later 11:15
   20 minutes later 11:05 45 minutes later 11:30

6.

a.		b.	
It is 1:50. → 40 minutes till		It is 7:20. → 40 minutes till	
It is 1:55. → 35 minutes till	2:30	It is 7:35. → 25 minutes till	8:00
It is 2:10. → 20 minutes till		It is 7:45. → 15 minutes till	
It is 2:25. → 5 minutes till		It is 7:50. → 10 minutes till	

More on Elapsed Time, p. 150

1.

from	10:30	8:30	1:40	5:45	3:20 AM
to	11:30	12:30	7:40	11:45	12:20 PM
elapsed time	1 hour	4 hours	6 hours	6 hours	9 hours

2.

from	1:25	2:00	3:05	7:30	5:10
to	1:55	2:15	3:25	7:50	5:50
elapsed time	30 min	15 min	20 min	20 min	40 min

from	2:00	7:05	8:25	6:40	11:15
to	2:35	7:35	8:50	6:55	11:40
elapsed time	35 minutes	30 minutes	25 minutes	15 minutes	25 minutes

3. a. She will arrive at 4:10.   b. She is at home at 5:00.   c. He spent 25 minutes doing math.

4. a. It took 20 minutes for the bus trip.
   b. Music class is 30 minutes long.
   c. Sergio started answering emails at 9:35.

5.

TIME NOW:	• 15 min later 8:00	• 2 hours later 9:45
	• 30 min later 8:15	• 5 hours later 12:45

6.

• 2 hours earlier 10:30	• 40 min earlier 11:50	TIME NOW: 
• 1 hour earlier 11:30	• 25 min earlier 12:05	

Practice, p. 152

1. a. 1:50, 10 till 2   b. 4:25, 25 minutes past 4   c. 8:55, 5 till 9   d. 11:05, 5 past 11
   e. 3:40, 20 till 4   f. 7:25, 25 past 7   g. 5:30, half past 5   h. 12:00, 12 o'clock

2. 2:45, 7:10, 9:25, 6:00

3. 10 minutes, 10 minutes, 25 minutes, 30 minutes, 15 minutes

4. 4 hours, 6 hours, 7 hours, 12 hours, 9 hours

Clock to the Minute, p. 153

1. a. 1:03   b. 1:24   c. 1:14   d. 1:11

2. a. 1:57   b. 2:47   c. 3:24   d. 4:38   e. 5:23   f. 8:17   g. 8:43
   h. 5:44   i. 12:41   j. 10:49   k. 9:29   l. 11:11   m. 4:26   n. 3:31
   o. 1:37   p. 12:01   q. 8:53   r. 10:19   s. 3:33   t. 2:58

3. a. 1:57; 2:07   b. 5:23; 5:33   c. 11:51; 12:01   d. 12:41; 12:51
   e. 10:49; 11:09   f. 8:17; 8:37   g. 11:11; 11:31   h. 10:09; 10:29
   i. 12:21; 12:36   j. 2:48; 3:03   k. 1:14; 1:29   l. 3:24; 3:39

Elapsed Time in Minutes, p. 156

1. a. 13 minutes   b. 37 minutes   c. 44 minutes   d. 57 minutes
   e. 27 minutes   f. 36 minutes   g. 14 minutes   h. 23 minutes

2. a. 19 minutes   b. 13 minutes   c. 36 minutes   d. 27 minutes
   e. 19 minutes   f. 31 minutes   g. 49 minutes   h. 28 minutes

3. a. 22 minutes   b. 49 minutes   c. 35 minutes   d. 44 minutes   e. 21 minutes   f. 54 minutes

4. a. At 4:52 she should take it out.   b. It is 1:46 PM.
   c. She should wake up at 5:34 AM.   d. The class started at 1:45 PM.

Using the Calendar, p. 158

1. a. November 2nd   b. December 5th   c. November 14th
   d. December 6th   e. December 17th   f. November 22nd
   g. January 23rd   h. December 10th   i. November 23rd

2. Camp started July 22nd; the 26th was the last day of camp.

3. December 7th.

4. 25 days till Jack's birthday, and 33 days till Mom's birthday.

5. Mary should return it at the latest on April 18th.

Mixed Review Chapter 4, p. 160

1. a. 25   b. 43   c. 29   d. 389   e. 561   f. 803

2. a. 14   b. 66   c. 49   d. 140

3. $89 - 17 = 72 + 72 = 144$ or $2 \times 89 - 2 \times 17 = 144$; They cost $144.

4. a. 27, 28, 0   b. 24, 24, 24   c. 56, 63, 32   d. 36, 60, 21

5. a. $7 \times 2 = 14$; He read 14 books.
   b. $3 \times 7 + 5 = 26$; He put 26 pencils in the pencil cases.

6. If there is anything in parentheses, do it first. Do the multiplications before additions or subtractions. Then, do the additions and subtractions from left to right. The first step is highlighted.

   a. $2 + \boxed{5 \times 2} = 12$ b. $5 \times \boxed{(1 + 1)} = 10$ c. $\boxed{(4 - 2)} \times 7 = 14$

7. a. $22 + ? = 141$ OR $141 - ? = 22$; Davy weighs 119 pounds.
   b. $275 - 48 = \underline{227}$; The cheaper washer costs $227.

8. $430 + 430 + 280 + 280 = 1{,}420$ meters approximately
   or $400 + 400 + 300 + 300 = 1{,}400$ meters approximately.

Review Chapter 4, p. 162

1. a. 11:51; 12:01   b. 8:43; 8:53   c. 4:57; 5:07   d. 1:14; 1:24

2. a. 15 min.   b. 35 min.   c. 38 min.   d. 11 min.

3. It ends at 2:00.

4. The trip was 3 hours long.

5. April 2nd.

6. February 27th.

Chapter 5: Money

Using the Half-Dollar, p. 165

1. a. 150 cents   b. 100 cents   c. 125 cents   d. 175 cents   e. 200 cents

2. a. 3 half-dollars   b. 4 half-dollars   c. 6 quarters
   d. 3 quarters   e. 3 half-dollars and 1 quarter   f. 4 half-dollars and 1 quarter

3. a. 81 cents   b. 75 cents   c. 100 cents   d. 80 cents   e. 176 cents
   f. 105 cents   g. 170 cents   h. 228 cents   i. 220 cents   j. 150 cents

Dollars, p. 167

1. a. $1.15   b. $5.16   c. $10.40   d. $6.26   e. $8.37   f. $11.56

2. a. $2.10   b. $6.54   c. $2.45

3. a. $0.80   b. $0.42   c. $0.25   d. $0.80   e. $0.12   f. $0.95

4. a. $0.56   b. $0.06   c. $4.25   d. 569¢   e. 30¢   f. 306¢

5. Mark has $7.45.

6. a. $4.04   b. $1.89   c. $6.39   d. $5.39   e. $1.23   f. $1.13

Making Change, p. 170

1. a. $0.24   b. $1.10   c. $2.65   d. $5.82   e. $6.96

2. a. $1.45   b. $1.40   c. $5.30   d. $2.01   e. $1.75   f. $5.85

3. a. $3.00   b. $16.00   c. $4.50   d. $7.60   e. $2.40   f. $3.70   g. $1.20   h. $0.60   i. $2.80

4. a. No, the correct change is $2.14.   b. No, the correct change is $2.24.

5. a. 44   b. 81   c. 28   d. 56   e. 66

6. a. 46¢, 24¢, 73¢   b. 62¢, 87¢, 14¢   c. 67¢, 61¢, 63¢

7. b. Change: $2.12. Use 2 dollars, a dime, and 2 pennies.
   c. Change: $2.24. Use 2 dollars, 2 dimes, and 4 pennies.
   d. Change: $1.05. Use 1 dollar, and 1 nickel or five pennies.
   e. Change: $0.74. Use 2 quarters, 2 dimes, and 4 pennies.
   f. Change: $3.72. Use 3 dollars, 2 quarters, 2 dimes, and 2 pennies.

Mental Math and Money Problems, p. 174

1. a. $4.10   b. $2.00   c. $2.30   d. $2.75   e. $38.10   f. $22.30
   g. $2.40   h. $1.05   i. $5.00   j. $3.30   k. $38.70   l. $3.65

2. a. $0.70, $0.50, $0.28   b. $0.70, $0.80, $0.38   c. $0.85, $0.44, $0.16

3. a. $3.20   b. $0.74   c. $2.81   d. $4.18

4. a. $0.55   b. $2.60   c. $0.35   d. $0.64   e. $0.22   f. $3.66   g. $0.21   h. $0.17

5. a. $2.90 total; $0.10 change   b. $3.45 total; $1.55 change
   c. Yes, I can, and my change is 20 cents.   d. No, I can't; I need 60 cents more.

Solving Money Problems, p. 177

1. a. $35.44   b. $16.90   c. $35.55   d. $107.08

2. a. $27.44   b. $10.97   c. $10.80

3. a. $7.35   b. $12.50   c. $5.55   d. $21.65

4. a. $6.45   b. $12.71   c. $3.56   d. $15.44

5. a. $6.20. The two mice cost $6.90 + $6.90 = $13.80. The difference of $20 and $13.80 is $6.20.
   b. Yes, it was correct. $35.90 + $14.10 = $50.
   c. Ernest can buy 2 calculators. $3.48 + $3.48 = $6.96, but $3.48 + $3.48 + $3.48 = $10.44
   d. He needs $4.73 more. A calculator and a book cost $3.48 + $6.75 = $10.23. Subtract to find how much more he needs: $10.23 − $5.50 = $4.73.

6. a. $17.94   b. $32.06
   c. Her change is $2.30. The total bill is $4.55 + $2.30 + $0.85 = $7.70. Change: $10 − $7.70 = $2.30
   d. John's total bill was $9.83, and his change was $10.17.
   e. $14.55 + $23.95 = $38.50. So, yes, she can, and her change is $1.50.

Mixed Review Chapter 5, p. 181

1. a.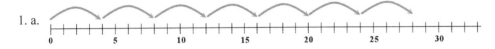

   b.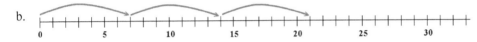

2. a. 50 − (20 − 7) = 37   b. (8 − 5) × 2 − 1 = 5   c. (15 + 5) × (2 − 1) = 20   OR   15 + 5 × (2 − 1) = 20

3. a. 72, 49, 54   b. 48, 35, 28   c. 36, 81, 72   d. 84, 64, 18

4. a. 5 × 8 = 40; You can buy eight pairs of socks.
   b. 7 × 4 = 28; There will be four layers of dominoes.

5.
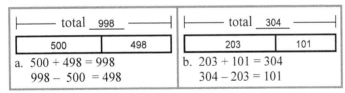
a. 500 + 498 = 998
   998 − 500 = 498
b. 203 + 101 = 304
   304 − 203 = 101

6. a. 3, 7, 9   b. 9, 6, 5   c. 8, 6, 4   d. 6, 9, 7

7. a. The tickets would cost $142.
   b. There are 52 pages left to read.
   c. There are 85 blue ribbons.

Review Chapter 5, p. 183

1. a. $10.40   b. $7.56

2. a. $0.75   b. $1.73   c. $1.45

3. a. Maria still needs to save $19.95.   b. Arnold's total bill is $6.70.   c. His change is $3.30.

4. a. My total bill is $3.55.   b. My change is $1.45.

Math Mammoth Grade 3-B Answer Key

By Maria Miller

Copyright 2007-2019 Maria Miller.

EDITION 6/2019

All rights reserved. No part of this book may be reproduced or transmitted in any form or by any means, electronic or mechanical, or by any information storage and retrieval system, without permission in writing from the author.

Copying permission: For having purchased this book, the copyright owner grants to the teacher-purchaser a limited permission to reproduce this material for use with his or her students. In other words, the teacher-purchaser MAY make copies of the pages, or an electronic copy of the PDF file, and provide them at no cost to the students he or she is actually teaching, but not to students of other teachers. This permission also extends to the spouse of the purchaser, for the purpose of providing copies for the children in the same family. Sharing the file with anyone else, whether via the Internet or other media, is strictly prohibited.

No permission is granted for resale of the material.

The copyright holder also grants permission to the purchaser to make electronic copies of the material for back-up purposes.

If you have other needs, such as licensing for a school or tutoring center, please contact the author at
https://www.MathMammoth.com/contact.php

Contents

Chapter 6: Place Value with Thousands

	Work-text page	Answer key page
Thousands	8	54
Four-Digit Numbers and Place Value	12	54
Which Number is Greater?	16	56
Mental Adding and Subtracting	18	56
Add 4-Digit Numbers with Regrouping	22	58
Subtract 4-Digit Numbers with Regrouping	24	58
Rounding to the Nearest Hundred	28	59
Estimating	31	60
Word Problems	34	61
Mixed Review Chapter 6	37	62
Review Chapter 6	39	62

Chapter 7: Geometry

	Work-text page	Answer key page
Shapes	46	63
Some Special Quadrilaterals	50	64
Perimeter	53	65
Problems with Perimeter	56	65
Getting Started with Area	59	66
More About Area	61	67
Multiplying by Whole Tens	65	68
Area Units and Problems	69	68
Area and Perimeter Problems	73	69
More Area and Perimeter Problems	75	69
Solids	78	71
Mixed Review Chapter 7	80	72
Geometry Review	82	72

Chapter 8: Measuring

	Work-text page	Answer key page
Measuring to the Nearest Fourth-Inch	87	73
Centimeters and Millimeters	91	73
Line Plots and More Measuring	95	74
Feet, Yards, and Miles	98	74
Meters and Kilometers	100	74
Pounds and Ounces	102	75
Grams and Kilograms	106	75
Cups, Pints, Quarts, and Gallons	110	76
Milliliters and Liters	113	76
Mixed Review Chapter 8	115	77
Review Chapter 8	117	77

Chapter 9: Division

	Work-text page	Answer key page
Division as Making Groups	122	79
Division and Multiplication	126	80
Division and Multiplication Facts	130	81
Dividing Evenly into Groups	133	81
Division Word Problems	137	82
Zero in Division	140	83
When Division is Not Exact	143	84
More Practice with the Remainder	146	84
Mixed Review Chapter 9	148	85
Review Chapter 9	150	86

Chapter 10: Fractions

	Work-text page	Answer key page
Understanding Fractions	155	87
Fractions on a Number Line	159	88
Mixed Numbers	163	90
Equivalent Fractions	167	91
Comparing Fractions 1	170	92
Comparing Fractions 2	173	94
Mixed Review Chapter 10	175	95
Fractions Review	177	95

Chapter 6: Place Value

Thousands, p. 8

1. a. 1312   b. 1130   c. 1057   d. 1502   e. 2330   f. 3478

2. a. 1256   b. 3594   c. 4617   d. 9822   e. 6211   f. 5799

3. a. 1001   b. 2005   c. 4061   d. 3012   e. 6200   f. 5090
   g. 1103   h. 7506   i. 5800   j. 2011   k. 2320   l. 9032

4. The numbers for the number lines are:

   1005 1006 1007 1008 1009 1010 1011

   1094 1095 1096 1097 1098 1099 1100

   1455 1456 1457 1458 1459 1460 1461

   1328 1329 1330 1331 1332 1333 1334

5.

1010	1020	1030	1040	1050
1060	1070	1080	1090	1100
1110	1120	1130	1140	1150
1160	1170	1180	1190	1200
1210	1220	1230	1240	1250

Four-Digit Numbers and Place Value, p. 12

1.

a. 1,034 = <u>1</u> thousand <u>0</u> hundreds <u>3</u> tens <u>4</u> ones   = 1000 + <u>0</u> + <u>3 0</u> + <u>4</u>
b. 5,670 = <u>5</u> thousand <u>6</u> hundreds <u>7</u> tens <u>0</u> ones   = 5000 + 600 + 70 + 0
c. 3,508 = <u>3</u> thousand <u>5</u> hundreds <u>0</u> tens <u>8</u> ones   = 3,000 + 500 + 0 + 8
d. 8,389 = <u>8</u> thousand <u>3</u> hundreds <u>8</u> tens <u>9</u> ones   = 8,000 + 300 + 80 + 9
e. 9,007 = <u>9</u> thousand <u>0</u> hundreds <u>0</u> tens <u>7</u> ones   = 9,000 + 0 + 0 + 7
f. 7,214 = <u>7</u> thousand <u>2</u> hundreds <u>1</u> tens <u>4</u> ones   = 7,000 + 200 + 10 + 4

Four-Digit Numbers and Place Value, cont.

2.

a. Five thousand nine hundred ninety	b. Six thousand sixteen	c. Six thousand three hundred three
T H T O 5 9 9 0	T H T O 6 0 1 6	T H T O 6 3 0 3
d. Eight thousand seven hundred	e. Nine thousand two hundred forty-five	f. Ten thousand
T H T O 8 7 0 0	T H T O 9 2 4 5	ten thou- T H T O sands 1 0 0 0 0

3. a. 2,090; 3,200   b. 8,005; 1,087   c. 8,220; 2,598   d. 4,050; 2,807

4. a. 3   b. 600   c. 50   d. 2,000

5. a. <u>6,048</u>, 6,049, <u>6,050</u>   b. <u>2,323</u>, 2,324, <u>2,325</u>   c. <u>1,799</u>, 1,800, <u>1,801</u>
   d. <u>8,808</u>, 8,809, <u>8,810</u>   e. <u>7,384</u>, 7,385, <u>7,386</u>   f. <u>9,243</u>, 9,244, <u>9,245</u>

6. a. 4,907   b. 8,586   c. 2,047   d. 4,620   e. 7,808   f. 5,060   g. 3,004   h. 9,500

7. a. 8,000   b. 6   c. 500   d. 40

8.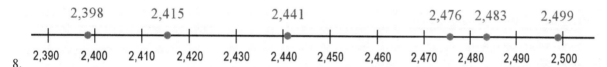

9.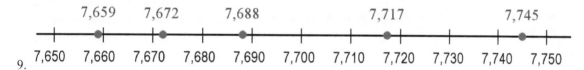

10. The number in the middle is 5,826.

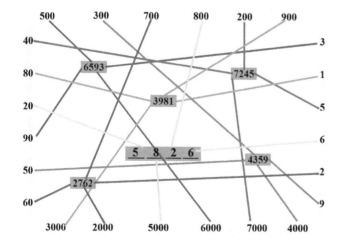

11. Solutions vary; the one below is just one example.

5000	+	200	+	3	+	3	= 5206
+		+		+		+	
11	+	3000	+	7	+	60	= 3078
+		+		+		+	
5	+	20	+	1000	+	900	= 1925
+		+		+		+	
6	+	15	+	398	+	13	= 432
=		=		=		=	
5022		3235		1408		976	

Which Number is Greater?, p. 16

1. a. 8,500   b. 5,700   c. 8,400   d. 3,500   e. 5,100   f. 2,770
   g. 3,811   h. 5,743   i. 7,011   j. 6,101   k. 9,834   l. 9,603

2.

a.	b.	c.	d.
1,050 < 5,095	220 < 1,020	1,307 > 1,032	4,012 < 4,284
2,400 < 2,750	8,060 > 6,999	4,906 < 6,029	5,008 < 5,040
6,005 > 4,500	1,007 < 1,705	5,077 < 5,570	1,890 < 1,897

3. a. 7,550   b. 2,338   c. 7,099   d. 1,212
   e. 8,502   f. 4,111   g. 1,809   h. 3,489

4. a. 700 + 50 $>$ 700 + 30 + 4   b. 500 + 6000 $=$ 6000 + 500
   c. 20 + 3000 $>$ 300 + 2000   d. 900 + 8 $<$ 9000 + 8
   e. 4000 + 80 $>$ 80 + 4 + 800   f. 30 + 6000 + 3 $>$ 300 + 60 + 3000
   g. 800 + 7000 + 2 $>$ 700 + 80 + 7000
   h. 500 + 3000 + 80 + 6 $=$ 6 + 80 + 500 + 3000

5. 3040 < 3899 < 4003 < 4203 < 4330

Mental Adding and Subtracting, p. 18

1. 5000, 5100, 5200, 5300, 5400, 5500
   2800, 2900, 3000, 3100, 3200, 3300

2.

a. ten hundreds = 1000	b. 56 hundreds = 5600
21 hundreds = 2100	79 hundreds = 7900
42 hundreds = 4200	80 hundreds = 8000

3.

a. 5000 + 200 = 5,200	b. 2900 + 200 = 3,100
5100 + 400 = 5,500	3100 + 300 = 3,400
c. 6800 + 400 = 7,200	d. 5600 − 200 = 5,400
3800 + 800 = 4,600	4500 − 300 = 4,200
e. 8000 − 200 = 7,800	f. 7900 − 800 = 7,100
8000 − 700 = 7,300	8500 − 700 = 7,800
g. 2200 − 600 = 1,600	h. 9800 − 700 = 9,100
3500 − 600 = 2,900	1300 − 300 = 1,000

4.

a. 600 + 400 = _1000_	b. 6600 + 400 = 7000
2500 + 500 = _3000_	2400 + 600 = 3000
c. 500 + 500 = 1000	d. 8200 + 800 = 9000
9200 + 800 = 10000	7300 + 700 = 8000

Mental Adding and Subtracting, cont.

5.

a. 5000 + 1200 = 6200 5100 + 2400 = 7500	b. 2700 + 3200 = 5900 3100 + 6300 = 9400
c. 2500 + 2500 = 5000 3500 + 3500 = 7000	d. 1600 + 1700 = 3300 3600 + 4500 = 8100

6. His trip was 3,500 km one way and 7,000 km both ways.

7. a. △ = 300   b. △ = 500   c. △ = 500   d. △ = 6,500   e. △ = 6,900   f. △ = 7,400

8. a. $3,700 + ? = $5,000   ? = $1,300; She still needs $1,300.

   b. 4,200 − 3,100 = ? OR 3,100 + ? = 4,200   ? = 1,100; The shortcut shortens the track by 1,100 feet.

   c. 3,100 + 3,100 + 3,100 = ?   ? = 9,300; He jogged a total of 9,300 feet.

   d. ? − $500 − $700 = $1,000   ? = $2,200; His paycheck is $2,200.

   e. $800 + $800 − ? = $1,200   ? = $400; He took $400 off the price.

9. a. 4000, 4010, 4020, 4030, 4040, 4050
   b. 1720, 1730, 1740, 1750, 1760, 1770
   c. 3350, 3360, 3370, 3380, 3390, 3400

10.

a. 100 + 20 = 120 5100 + 20 = 5120	b. 220 + 40 = 260 4220 + 40 = 4260
c. 140 − 90 = 50 4140 − 90 = 4050	d. 230 − 30 = 200 4230 − 30 = 4200

11.

a. 4980 + 20 = 5000 980 + 20 = 1000	b. 7210 + 90 = 7300 210 + 90 = 300
c. 7760 − 30 = 7730 760 − 30 = 730	d. 5540 + 50 = 5590 540 + 50 = 590

Puzzle corner. The puzzle has MANY possible solutions. Basically you just pick one number at will and start filling the puzzle in, and if you run into a difficulty, you change the number. This is just an example solution.

4550	−	14	+	24	=	4560
−		+		−		
0	+	30	+	20	=	50
+		−		+		
30	+	14	+	56	=	100
=		=		=		
4580		30		60		

Add 4-Digit Numbers with Regrouping, p. 22

1. a. 5601 b. 7109 c. 7672 d. 7386 e. 4770 f. 6818 g. 9472 h. 8162 i. 9277

2. a. 6,293   b. 4,668

3. a. The total is $4,594.   b. The total cost is $2,220.

Puzzle corner:

```
   3  [9]  5  [2]          2  9  [8] [1]
+ [5]  1  [2]  9        + [2][4]  3   6
  ─────────────            ─────────────
   9   0   8   1           5  4   1   7
```

Subtract 4-Digit Numbers with Regrouping, p. 24

1. a. 4,581   Check: 4,581 + 510 = 5,091   b. 1,197   Check: 1,197 + 1,716 = 2,913
   c. 7,024   Check: 7,024 + 1,378 = 8,402   d. 5,970   Check: 5,970 + 911 = 6,881
   e. 3,056   Check: 3,056 + 3,490 = 6,546   f. 4,055   Check: 4,055 + 5,025 = 9,080
   g. 3,393   Check: 3,393 + 1,116 = 4,509   h. 4,144   Check: 4,144 + 2,065 = 6,209

2. a. 1,786   b. 2,276   c. 295   d. 6,099   e. 2,523   f. 4,926   g. 1,899   h. 6,226

3. a. 3,285   b. 4,957   c. 4,237   d. ▲ = 6,507

4. a.

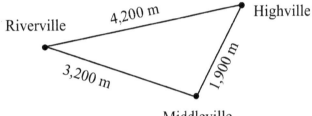

   b. The total distance is 9,300 m.
   c. 2,600 m more

Puzzle corner:

```
   3 [8] 5 [3]      8  9 [5]  4      6  [0] [0]  9     [7]  0   0   3
 -[2] 1 [8] 9    - [4] 2  3   6    -  3   2   2   5    - 2   8  [6][7]
   ─────────        ─────────────     ─────────────      ─────────────
   1  6  6  4       4  7  1   8      2   7   8   4       4  1   3   6
```

Rounding to the Nearest Hundred, p. 28

1. a. 800, 900   b. 800, 900   c. 900, 800   d. 800, 900

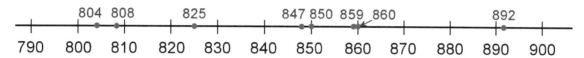

2. a. 400, 800   b. 500, 1,000   c. 700, 600   d. 300, 100

3. a. 2,400   b. 2,400   c. 2,500   d. 2,500

4. a. 6,200   b. 5,900   c. 1,700   d. 8,300   e. 8,000   f. 2,200
   g. 1,800   h. 6,800   i. 800   j. 300   k. 9,600   l. 3,500

5. a. 7,000   b. 6,000   c. 2,000   d. 4,900   e. 2,900   f. 10,000   g. 1,000   h. 10,000   i. 800

6. Across:
   a. 2,264 ≈ 2,300
   b. 4,973 ≈ 5,000
   c. 4,248 ≈ 4,200
   d. 545 ≈ 500

   Down:
   e. 3,709 ≈ 3,700
   f. 672 ≈ 700
   g. 5,370 ≈ 5,400
   h. 8,816 ≈ 8,800

a. 2	e. 3	0	0	
	7			
b. 5	0	0	0	
	0			h. 8
f. 7		g. 5		8
0		c. 4	2	0	0
0		0		0
	d. 5	0	0	

7. a. Usually, Mary receives about 100 spam emails daily, but on 5/9 she got about 300 spams.

   b. During the work week from 5/7 till 5/11 she received about 700 spams.

Spam Emails Mary Received		
Date	Spams	round to nearest 100
Mo 5/7	125	100
Tu 5/8	97	100
Wd 5/9	316	300
Th 5/10	118	100
Fr 5/11	106	100

Estimating, p. 31

1.

a. Estimate:	569	+	234		Calculate exactly:		569 + 234 803
	↓		↓				
	600	+	200	= 800			
b. Estimate:	8,155	+	424		Calculate exactly:		8155 + 424 8579
	↓		↓				
	8,200	+	400	= 8,600			
c. Estimate:	577	−	125		Calculate exactly:		577 − 125 452
	↓		↓				
	600	−	100	= 500			
d. Estimate:	7,028	−	465		Calculate exactly:		7028 − 465 6563
	↓		↓				
	7,000	−	500	= 6,500			

2.

a. Estimate:	5,171	+	568				Calculate exactly:		5171 + 568 5739
	↓		↓						
	5,200	+	600	= 5,800					
b. Estimate:	4,162	+	3,439				Calculate exactly:		4162 + 3439 7601
	↓		↓						
	4,200	+	3,400	= 7,600					
c. Estimate:	7,577	−	2,947				Calculate exactly:		577 − 2947 4630
	↓		↓						
	7,600	−	2,900	= 4,700					
d. Estimate:	756	+	4,178	+	836		Calculate exactly:		756 4178 + 836 5770
	↓		↓		↓				
	800	+	4,200	+	800	= 5,800			
e. Estimate:	8,295	−	5,538	−	1,150	−	924		
	↓		↓		↓		↓		
	8,300	−	5,500	−	1,200	−	900	= 700	

Calculate exactly:	8295 − 5538 2757	2757 − 1150 1607	1607 − 924 683

Instead of three subtractions, you could also add 5,538 + 1,150 + 924, and then subtract that sum from 8,295.

Now check. Were your estimations close to the real answers?
Yes.

Estimating, cont.

3.

a. Here, Elisa makes an error in regrouping.	Elisa's work:	6 5 4 0 − 2 5 9 ───── 6 <u>3 9</u> 1	Correct answer:	13 4 3̶ 10 6 5̶ 4̶ 0̶ − 2 5 9 ───── 6 2 8 1
b. Elisa adds instead of subtracting in the hundreds.	Elisa's work:	3 8 3 4 − 1 2 6 0 ───── 2 <u>9</u> 7 4	Correct answer:	7 13 3 8̶ 3̶ 4 − 1 2 6 0 ───── 2 5 7 4
c. Elisa forgets to add the regrouped 1 thousand.	Elisa's work:	3 8 7 4 + 1 9 9 0 ───── <u>4</u> 8 6 4	Correct answer:	1 1 3 8 7 4 + 1 9 9 0 ───── 5 8 6 4

4. Yes, the estimation was close.

		Rounded numbers:
rice 1 kg	2750	2,800
parsley	449	400
potatoes	1876	1,900
tomatoes	1564	1,600
bananas	1238	1,200
onions	946	900
TOTAL	8823	8,800

Word Problems, p. 34

1. Estimate: $1,200 + $700 = $1,900; change $100. Exact: $2,000 − $1903 = $97

2. Estimate: $8,700 − $1,300 = $7,400. Exact: $8,740 − $1,295 = $7,445.

3. 4,321 − 1,234 = 3,087

4. Yes, you can. You can estimate: $979 is close to $1,000, so you can buy three of them for $3,000.
   The total cost is $979 + $979 + $979 = $2,937. Change is $3,000 − $2,937 = $63.
   Or, you can solve the change this way: $979 is $21 less than $1,000, so three of them cost 3 × $21 = $63 less than $3,000.

5. Estimate: $1,100 + $1,100 + $1,100 + $1,100 − $500 = $3,900.
   Exact: The total bill is $1,109 + $1,109 + $1,109 + $1,109 − $500 = $3,936.

6. a. Wednesday he caught 1,300 kg of fish.
   b. Friday, he caught 1,100 kg of fish.
   c. He caught 500 kg more.
   d. He caught 3,700 kg of fish during the week.

Word Problems, cont.

7. a.
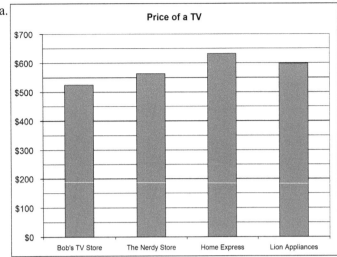

b. The difference is $632 − $525 = $107.

Puzzle corner:

```
  5 [2][3] 7        [6][3] 0 1       6 [7][6] 9      [2][1] 8 8
− [1] 5 5 [3]     −  2 7 [5][9]    + [3] 0 0 5     + 2 7 [1][2]
  ─────────        ─────────       ─────────       ─────────
  3 6 8 4            3 5 4 2         9 7 7 4         4 9 0 0
```

Mixed Review Chapter 6, p. 37

1. a. 49, 249   b. 63, 663   c. 75, 275

2. a. 6 × 2 + 4 × 4 = 28 legs   b. 9 × 4 + 3 = 39 windows

3. a. 570, 340   b. 900, 260   c. 600, 430   d. 60, 1,000

4. a. 3, 8, 14   b. 19, 24, 60   c. 85, 53, 40   d. 43, 125, 271

5. a. XV, XIX   b. XXI, XLIII   c. LVI, LXV   d. XC, XCIX

6. a. 4, 9, 7   b. 7, 5, 8   c. 8, 6, 4   d. 4, 8, 5

7. a. 6 hours   b. 5 hours   c. 20 minutes   d. 47 minutes

8. a. $4.40   b. $15.27   c. $27.64

Review Chapter 6, p. 39

1. a. 7,240   b. 6,005   c. 2,029

2. a. 7,503; 3,090   b. 1,037; 6,400

3. a. >   b. >   c. <   d. <

4. 1,900; 7,200   b. 3,300; 3,700   c. 800; 900   d. 4,900; 8,300

5. a. △ = 700   b. △ = 9,800   c. △ = 1,500

6. a. 900   b. 5,300   c. 6,000   d. 2,700

7. a. Estimate: 2,500 + 1,800 = 4,300, Exact: 4,343   b. Estimate: 6,600 − 700 = 5,900, Exact: 5,845

8. a. Estimate: $1,600 + $300 + 1,000 = $2,900.   Exact: $2,863.
   b. Estimate: $5,000 − $300 − $1,300 = $3,400.   Exact: $5,000 − $278 − $1,250 = $3,472.

Chapter 7: Geometry

Shapes, p. 46

1. Answers vary. For example: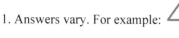

2. a. a pentagon   b. a quadrilateral (a kite)   c. a pentagon   d. a hexagon

3. Answers can vary. These are example answers.

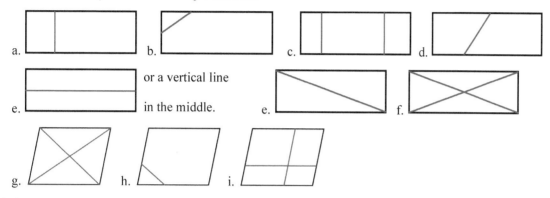

4. Answers vary.

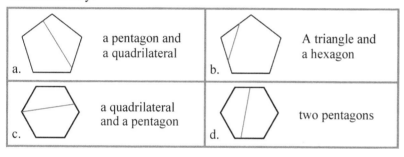

5. a. rectangle   b. quadrilateral   c. octagon and a square

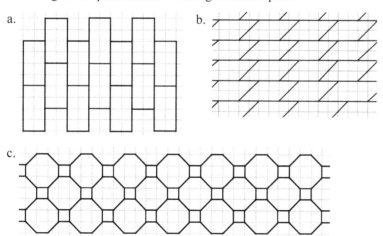

Some Special Quadrilaterals, p. 50

1. Answers vary. Here are some examples:

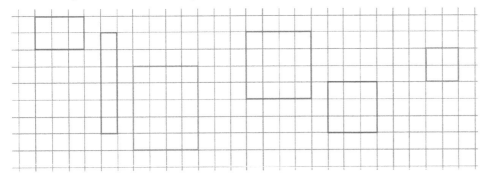

2.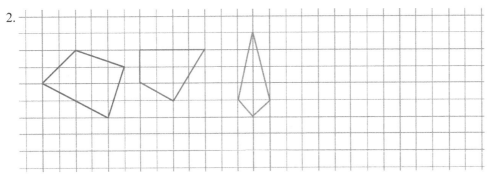

4. a. a square   b. a rhombus   c. a square   d. a rhombus

5. Yes, a square is a rhombus, because all of its four sides have the same length.

6.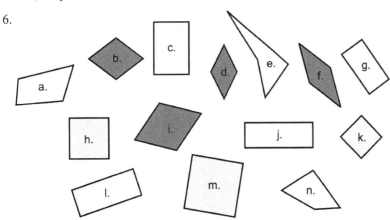

6. The rectangles are c, g, j, and l. The squares are h, k, and m. The rhombi are b, d, f, and i. Other quadrilaterals are a, e, and n.

7.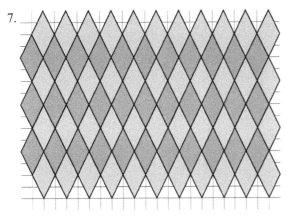

Perimeter, p. 53

1. a. 14 units   b. 12 units   c. 12 units
   d. 12 units   e. 18 units   f. 24 units

2. a. 16 cm   b. 16 cm   c. 12 cm + 5 cm + 13 cm = 30 cm

3. a. 6 in.   b. 10 in.

To find the perimeter, simply **add all the side lengths.** How many units is the perimeter of the triangle on the right? It is 8 + 9 + 10 units, or __27__ units.	(triangle with sides 8, 9, 10)
Often you need to figure out some side lengths that are not given. What side lengths are not given? The perimeter is __24__ cm.	(rectangle 8 cm × 4 cm)

4. a. 24 units   b. 48 units   c. 3 in.   d. 42 cm   e. 24 cm   f. 11 in.

5. a. 28 in.   b. 52 cm

Problems with Perimeter, p. 56

1. a. 6 + _?_ + 6 + _?_ = 20   or   6 + _?_ = 10. The unknown _?_ = 4 cm

   b. 15 + _?_ + 15 + _?_ = 44   or   15 + _?_ = 22. The unknown _?_ = 7 cm

   c. 12 + _?_ + 12 + _?_ = 82   or   12 + _?_ = 41. The unknown _?_ = 29 in.

   d. _?_ + _?_ + _?_ + _?_ = 12   or   4 × _?_ = 12. The unknown _?_ = 3 in.

2. a. _?_ + _?_ + _?_ + _?_ = 44   or   4 × _?_ = 44. The unknown _?_ = 11 cm.

   b. The perimeter is 48 in.

   c. P = 12 cm + 4 cm + 8 cm + 6 cm + 4 cm + 10 cm = 44 cm

3. Just counting the units in the picture, the perimeter is 18 units. Since each unit is 10 feet, we get 18 × 10 feet = 180 feet. Or, you can count by tens as you count the units for the perimeter.

4. 120 feet

5. 6 m

Problems with Perimeter, cont.

6. Answers vary. In each rectangle, the two side lengths should add up to 12 units (half of the perimeter).

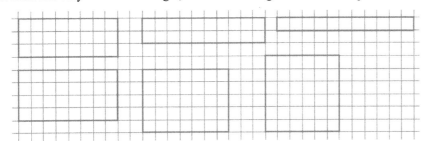

One side	Other side	Perimeter
3 units	9 units	24 units
1 unit	11 units	24 units
2 units	10 units	24 units
4 units	8 units	24 units
5 units	7 units	24 units
6 units	6 units	24 units

Puzzle corner: Answers vary, for example:

8 units: 10 units: 14 units:

Getting Started with Area, p. 59

1. a. 8 square units   b. 13 square units   c. 8 square units   d. 12 square units

2.

a. 2 × 5 = 10 A = 10 square units.	b. 3 × 3 = 9 A = 9 square units.	c. 6 × 3 = 18 A = 18 square units.

3. a. 15 square units   b. 12 square units   c. 10 square units   d. 17 square units

4. a. 32 square units   b. 31 square units

5. The rectangles can be 1 × 16, 2 × 8, or 4 × 4.

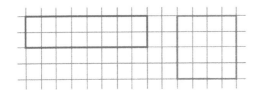

6. The rectangles can be 1 × 24, 2 × 12, 3 × 8, or 4 × 6.

66

More About Area, p. 61

1. a. $3 \times 3 + 3 \times 5 = 24$ b. $2 \times 5 + 3 \times 3 = 19$
   c. $3 \times 5 + 2 \times 3 = 21$ d. $4 \times 5 + 2 \times 4 = 28$

2. a. $4 \times (2 + 5) = 4 \times 2 + 4 \times 5$

   b. $4 \times (4 + 2) = 4 \times 4 + 4 \times 2$

   c. $5 \times (3 + 4) = 5 \times 3 + 5 \times 4$

   d. $3 \times (4 + 2) = 3 \times 4 + 3 \times 2$

   e. $2 \times (3 + 3) = 2 \times 3 + 2 \times 3$

3.

a.	$3 \times (2 + 4)$ =	3×2 +	3×4	
	area of the whole rectangle	area of the first part	area of the second part	
b.	$5 \times (1 + 4)$ =	5×1 +	5×4	
	area of the whole rectangle	area of the first part	area of the second part	
c.	$4 \times (3 + 1)$ =	4×3 +	4×1	
	area of the whole rectangle	area of the first part	area of the second part	
d.	$3 \times (2 + 1)$ =	3×2 +	3×1	
	area of the whole rectangle	area of the first part	area of the second part	
e.	$2 \times (5 + 2)$ =	2×5 +	2×2	
	area of the whole rectangle	area of the first part	area of the second part	

4. a. $3 \times 3 + 3 \times 6 + 3 \times 4 = 39$ square units
   b. $6 \times 8 - 3 \times 3 = 39$ square units
   c. $7 \times 4 + 5 \times 3 + 7 \times 4 = 71$ square units or $13 \times 7 - 5 \times 4 = 71$ square units

Puzzle corner. $3 \times 4 + 4 \times 6 - 4 \times 1 = 32$ squares

Multiplying by Whole Tens, p. 65

1.

9 × 10 = 90	14 × 10 = 140	19 × 10 = 190
10 × 10 = 100	15 × 10 = 150	20 × 10 = 200
11 × 10 = 110	16 × 10 = 160	21 × 10 = 210
12 × 10 = 120	17 × 10 = 170	22 × 10 = 220
13 × 10 = 130	18 × 10 = 180	23 × 10 = 230

There is a pattern: *Every answer ends in 0.* Also, there is something special about the number you multiply times 10, and the answer. Can you see that? <u>You simply add a zero on the end of the number.</u>

2. a. 110, 560   b. 990, 180   c. 820, 0

3. a. 50, 500   b. 900, 900   c. 170, 17

4. a. Parts: 8 × 10 and 8 × 10.
   The total area is 160.
   b. Parts: 5 × 10 and 5 × 10 and 5 × 10.
   The total area is 150.
   c. Parts: 7 × 10 and 7 × 10 and 7 × 10.
   The total area is 210.
   d. Parts: 4 × 10 and 4 × 10 and 4 × 10 and 4 × 10.
   The total area is 160.

5. a. 3 × 40 = 40 + 40 + 40 = 120
   b. 2 × 80 = 80 + 80 = 160
   c. 4 × 40 = 40 + 40 + 40 + 40 = 160
   d. 5 × 30 = 30 + 30 + 30 + 30 + 30 = 150
   e. 5 × 70 = 70 + 70 + 70 + 70 + 70 = 350
   f. 3 × 80 = 80 + 80 + 80 = 240
   Multiply the numbers, then tack on the zero.

6.

a.	7 × 90 = <u>7</u> × <u>9</u> × 10 = <u>63</u> × 10 = 630	b.	4 × 80 = 4 × 8 × 10 = 32 × 10 = 320
c.	6 × 40 = 6 × 4 × 10 = 24 × 10 = 240	d.	9 × 90 = 9 × 9 × 10 = 81 × 10 = 810
e.	30 × 6 = 10 × 3 × 6 = 10 × 18 = 180	f.	80 × 3 = 10 × 8 × 3 = 10 × 24 = 240

7. a. 490   b. 480   c. 280
   d. 200   e. 210   f. 270

8. The area is 7 × 80 = 560 square units.

9. 7 × 10 = 70 square units

10. The total area: 8 × 30 = 240 square units.
    Area of each part: 8 × 10 = 80 square units.

11. The rectangle is divided into thirds. Each third has the area of 7 × 40 = 280 square units. The total area is then 280 + 280 + 280 = 840 square units.

Puzzle corner. Answers may vary. You can add 16 repeatedly: 16 + 16 + 16 + 16 + 16 = 80 squares. Or, you could divide the rectangle into two parts, each having the area of 5 × 8 = 40. Then the total area is 80 squares.

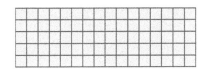

Area Units and Problems, p. 69

1. a. A = 2 cm × 4 cm = 8 cm²
   b. A = 6 cm × 3 cm = 18 cm²
   c. A = 8 cm × 2 cm = 16 cm²
   d. A = 4 cm × 3 cm = 12 cm²

2. a. A = 3 in. × 3 in. = 9 in²
   b. A = 2 in. × 4 in. = 8 in²
   c. A = 5 in. × 1 in. = 5 in²

3. a. A = 4 m × 3 m = 12 m²
   b. A = 5 ft × 6 ft = 30 ft²
   c. A = 12 cm × 4 cm = 48 cm²
   d. A = 8 in. × 7 in. = 56 in²

4. A = 11 m × 4 m + 4 m × 4 m = 60 m²

5. A = 4 ft × 6 ft + 12 ft × 6 ft = 96 ft²

6. Danny's room is 16 m². Joe's room is 15 m².
   Danny's room is bigger by one square meter.

7. The white square has the area of 3 in. × 2 in. = 6 in².
   The pink area is 6 in. × 8 in. − 3 in. × 2 in. = 42 in².

Area and Perimeter Problems, p. 73

1. a. perimeter 14 m; area 10 m^2

   b. perimeter 24 ft; area 36 ft^2

   c. perimeter 12 in.; area 8 in^2

   d. perimeter 12 cm; area 9 cm^2

2. a. You can divide the shape into four 4 m by 4 m squares, each having the area of 16 m^2. The area is then
   16 m^2 + 16 m^2 + 16 m^2 + 16 m^2 = 64 m^2.
   The perimeter is 40 m.

3. For the area, divide the shape into two rectangles. That can be done in two ways.
   You could get 11 cm × 4 cm + 4 cm × 8 cm = 76 cm^2.
   or 4 cm × 12 cm + 7 cm × 4 cm = 76 cm^2.

   The perimeter is
   4 cm + 8 cm + 7 cm + 4 cm + 11 cm + 12 cm = 46 cm.

4. a. 5 m × 4 m = 20 m^2 and 10 m × 4 m = 40 m^2.
   b. 60 m^2.
   c. 38 m

5. Area of each little part is 6 m × 10 m = 60 m^2.
   The total area is 6 m × 60 m = 360 m^2.

Puzzle corner. a. 13 × 3 rectangle.

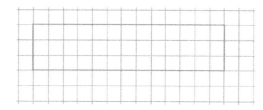

b. a 14 × 4 rectangle.

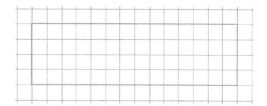

More Area and Perimeter Problems, p. 75

1. a. 20 m × 9 m = 180 m^2 and 30 m × 9 m = 270 m^2
   b. 450 m^2
   c. 118 m

2.

	first side	second side	area
Rectangle 1	2 units	12 units	24 square units
Rectangle 2	3 units	8 units	24 square units
Rectangle 3	4 units	6 units	24 square units
Rectangle 4	1 unit	24 units	24 square units

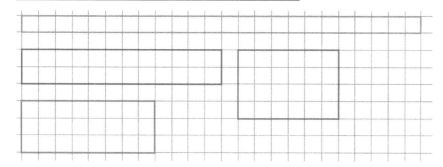

More Area and Perimeter Problems, cont.

3.

	one side	second side	area	perimeter
Rectangle 1	2 units	12 units	24 square units	28 units
Rectangle 2	3 units	8 units	24 square units	22 units
Rectangle 3	4 units	6 units	24 square units	20 units
Rectangle 4	1 unit	24 units	24 square units	50 units

4.

	first side	second side	perimeter
Rectangle 1	2 units	8 units	20 units
Rectangle 2	3 units	7 units	20 units
Rectangle 3	4 units	6 units	20 units
Rectangle 4	5 units	5 units	20 units
Rectangle 5	1 unit	9 units	20 units

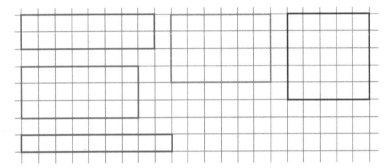

5.

	first side	second side	perimeter	area
Rectangle 1	2 units	8 units	20 units	16 square units
Rectangle 2	3 units	7 units	20 units	21 square units
Rectangle 3	4 units	6 units	20 units	24 square units
Rectangle 4	5 units	5 units	20 units	25 square units
Rectangle 5	1 unit	9 units	20 units	9 square units

6. a. 30 m × 9 m = 270 m² and 30 m × 6 m = 180 m²
   b. 450 m²
   c. 90 m

More Area and Perimeter Problems, cont.

7.

a. $3 \times (5 + 2) = 3 \times 5 + 3 \times 2$	
b. $4 \times (3 + 5) = 4 \times 3 + 4 \times 5$	
c. $2 \times (5 + 2) = 2 \times 5 + 2 \times 2$	
d. $3 \times (2 + 1) = 3 \times 2 + 3 \times 1$	

Puzzle corner. a. $9 \times (20 + 30) = 9 \times 20 + 9 \times 30$

b. $10 \times 20 = 200$ m^2

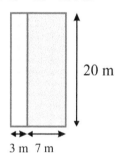

3 m 7 m

Solids, p. 78

1.

Shape	Name	E	V	F
a.	pyramid	8	5	5
b.	cube	12	8	6
c.	tetrahedron	6	4	4

2. a. Answers will vary. Please check the student's answer. For example: ice cream cone, safety cone, a steeple on a building.

   b. Answers will vary. Please check the student's answer. For example: a can, a roll of paper towels, bucket, a glass.

3.

Name of Figure	Faces	Edges	Vertices
rectangular prism	6	12	8
rectangular pyramid	5	8	5
tetrahedron	4	6	4

Mixed Review Chapter 7, p. 80

1. a. 2:59   b. 4:56   c. 9:09   d. 11:31

2. a. 293   b. 466   c. 2486   d. 2,162

3.

a. 99 + ☐ = 145	b. 34 + ☐ = 76
145 − 99 = 46	76 − 34 = 42

4. a. <   b. >   c. <   d. >   e. <   f. <

5. a. The volleyball set costs $24.
   b. The snorkeling set costs $4 more than the swim ring set.
   c. They cost $13.
   d. The cheapest would be $13.

6. a. change is $0.45   b. change is $4.12   c. change is $3.30

7.

a. 10 − (40 − 30) = 0	b. (4 + 5) × 2 − 1 = 17	c. 5 × (7 − 3) − 1 = 19

Geometry Review, p. 82

1. a. A, B, F, H, J   b. C, E, I, K, L

2. Answers vary. Check the student's answers.

3.

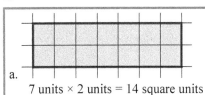

a. 7 units × 2 units = 14 square units
b. 4 × 5 = 20 square units

4. a. Area 35 cm²   b. perimeter 24 cm

5. a. Area 12 square units; perimeter 14 units   b. Area 11 square units; perimeter 24 units

6. a. A = 3 × 2 + 3 × 4 = 18 square units   b. A = 2 × 2 + 3 × 4 = 16 square units

7. a. 490   b. 480   c. 280

8. Area of each part: 9 × 10 = 90 square units. Total area 9 × 40 = 360 square units.

9.

a. 3 × (5 + 1) = 3 × 5 + 3 × 1

b. 4 × (2 + 3) = 4 × 2 + 4 × 3

Chapter 8: Measuring

Measuring to the Nearest Fourth-Inch, p. 87

1. a. 1 1/4 inches   b. 1 3/4 inches   c. 3 1/4 inches   d. 4 3/4 inches
   e. 5 1/2 inches   f. 4 1/4 inches   g. 3 3/4 inches

2. Check the student's answers. The answers below may not be the right length when printed from the download version, because many printers will print with "shrink to fit" or "fit to printable area."
   a. ──
   b. ──────────────────────
   c. ──
   d. ──────────────────────────────

3. Answers will vary.

4. The images are not to scale.

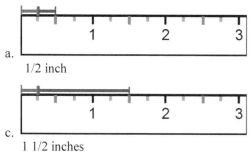
a. 1/2 inch

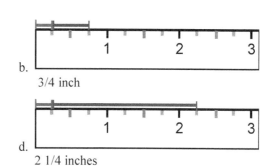
b. 3/4 inch

c. 1 1/2 inches

d. 2 1/4 inches

5. a. 1/2 in; 1 1/2 in   b. 1 in; 4 1/2 in   c. 6 in; 8 in   d. 3/4 in; 2 3/4 in

Centimeters and Millimeters, p. 91

1. a. 3 cm 4 mm = 34 mm   b. 7 cm 7 mm = 77 mm   c. 11 cm 6 mm = 116 mm
   d. 12 cm 9 mm = 129 mm   e. 6 cm 1 mm = 61 mm   f. 5 cm 3 mm = 53 mm

2. Check the student's answers. The answers below may not be the right length when printed from the download version, because many printers will print with "shrink to fit" or "fit to printable area."
   a. ──────────────────────────────────
   b. ────────────────────────
   c. ─────
   d. ──────────────
   e. ────────────────────────────────

3. Answers will vary. Check the student's work.

4. a. 10 mm; 20 mm; 50 mm; 80 mm   b. 11 mm; 12 mm; 18 mm; 23 mm
   c. 45 mm; 25 mm; 78 mm; 104 mm

5. a. 7 cm; 10 cm   b. 1 cm 2 mm; 4 cm 5 mm   c. 8 cm 9 mm; 10 cm 2 mm

6. The answers below may not match what you measure, if you have printed from the download version, because many printers will print with "shrink to fit" or "fit to printable area."
   side AB 53 or 54 mm   side BC 110 mm   side CA 117 mm

7. 280 or 281 mm

8. The answers below may not match what you measure, if you have printed from the download version, because many printers will print with "shrink to fit" or "fit to printable area."
   The sides measure 22 mm, 65 mm, and 79 mm. The perimeter is 166 mm or 16 cm 6 mm.

9. a. 20 mm   b. 98 mm   c. 63 mm   d. 69 mm   e. 76 mm   f. 85 mm   g. 46 mm   h. 641 mm

Line Plots and More Measuring, p. 95

1. a. 3 1/4 in.   b. 4 3/4 in.   c. 1 3/4 in.

2. Answers will vary. Please check the student's work.

3. Answers will vary. Please check the student's work.

4. The answers below may not match what you measure, if you have printed from the download version, because many printers will print with "shrink to fit" or "fit to printable area."

   Side AB 5 cm 7 mm
   Side BC 4 cm 7 mm
   Side CD 4 cm 5 mm
   Side DA 4 cm 5 mm
   Perimeter 19 cm 4 mm

5. Answers will vary.

6. Answers will vary.

Feet, Yards, and Miles, p. 98

1. Answers vary.

2. inch, foot, yard, mile

3. Answers vary.

4. a. mi   b. in   c. ft   d. in   e. ft   f. mi

5. perimeter = 30 ft + 30 ft + 10 ft + 10 ft = 80 ft; area = 10 ft × 30 ft = 300 ft^2

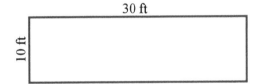

6.

a. 1 ft = 12 in 3 ft = 36 in 5 ft = 60 in	b. 1 ft 2 in. = 14 in 1 ft 8 in. = 20 in 1 ft 11 in. = 23 in	c. 2 ft 4 in. = 28 in 2 ft 6 in. = 30 in 3 ft 3 in. = 39 in

7. Emma is 50 inches tall.

8. Rebecca is three inches taller.

9. His train of pencils is two feet long.

10. The perimeter of the rectangle is 11 ft 8 in.

Meters and Kilometers, p. 100

1. Answers vary.

2. Answers will vary.

3. a. m   b. mm   c. cm   d. km   e. m   f. cm

4. Answers will vary.

5. mm, cm, m, km

Meters and Kilometers, cont.

6.

| a. 1 m = 100 cm
2 m = 200 cm
5 m = 500 cm | b. 1 m 20 cm = 120 cm
1 m 14 cm = 114 cm
1 m 58 cm = 158 cm | c. 5 m 85 cm = 585 cm
2 m 17 cm = 217 cm
3 m 8 cm = 308 cm |

7. The train of pillows would be two meters long.

8. Ellie is 7 cm taller.

9. The perimeter is 5 m 60 cm.

Pounds and Ounces, p. 102

1. a. 4 lb 0oz   b. 0 lb 9 oz   c. 1 lb 1 oz

2. a. 1 lb 11oz   b. 2 lb 4 oz   c. 5 lb 3 oz   d. 3 lb 9 oz   e. 3 lb 14 oz   f. 2 lb 15 oz   g. 5 lb 9 oz   h. 3 lb 8 oz   i. 2 lb 8 oz

3. Answers will vary.

4. Answers will vary.

5. Answers will vary.

6. a. 2 lb   b. 160 lb   c. 2 oz   d. 70 lb   e. 600 lb   f. 6 oz

7. a. lb   b. lb   c. oz   d. oz   e. lb

8.

| a. 2 lb = 32 oz
3 lb = 48 oz
4 lb = 64 oz | b. 1 lb 1 oz = 17 oz
1 lb 7 oz = 23 oz
2 lb 11 oz = 43 oz | c. 2 lb 4 oz = 36 oz
3 lb 9 oz = 57 oz
5 lb 4 oz = 84 oz |

9. The liquid weighs five ounces

10. a. 8 letters   b. 1 lb 14 oz

Grams and Kilograms, p. 106

1. a. 2 kg 200 g   b. 0 kg 200 g   c. 1 kg 400 g   d. 0 kg 800 g   e. 3 kg 0g   f. 3 kg 400 g

2. Answers will vary.

3. Answers will vary.

4. Answers will vary.

5. a. 5 g   b. 70 kg   c. 1 kg   d. 1 kg   e. 30 g   f. 15 g
   g. 2,000 kg   h. 100 g   i. 8 kg   j. 10 g   k. 5 kg   l. 300 g

6. An adult woman - 55 kg;   A puppy - 1 kg ;   A pencil - 50 g;
   A school book - 500 g;   A magazine - 150 g;   A 9-year-old boy - 25 kg

7. a. kg   b. g   c. kg   d. g   e. kg   f. kg   g. g

8.

| a. 1 kg = 1,000 g
2 kg = 2,000 g
3 kg = 3,000 g
4 kg = 4,000 g | b. 1 kg 600 g = 1,600 g
1 kg 80 g = 1,080 g
2 kg 450 g = 2,450 g
8 kg 394 g = 8,394 g | c. 9 kg = 9,000 g
8 kg 600 g = 8,600 g
5 kg 8 g = 5,008 g
7 kg 41 g = 7,041 g |

Grams and Kilograms, cont.

9. Five t-shirts would weigh 1 kg.

10. Their combined weight is 3 kg.

11. a. Their total weight is 2,100 grams.
    b. Their total weight is 2 kg 100 g.

12. The remaining potatoes weigh 1kg 900 g.

13. a. The total weight of the packages he has received this month is 8 kg 300 g.
    b. He is still allowed to receive 1 kg 700 g of mail this month.

Cups, Pints, Quarts, and Gallons, p. 110

1. It takes 2 pints of water to fill 1 quart jar.

2. How much water is left in the quart jar? 1 pint.

3. Two times. One pint is two cups.

4. Four times. One quart is four cups.

5. Answers will vary.

6. Answers will vary.

7. Answers will vary.

8. a. 2 pt   b. 4 C   c. 2 C

9. a. a pint is more.   b. two cups are equal to a pint.   c. 3 cups are more
   d. a quart is more   e. 1 quart is more   f. 2 pints are equal to 1 quart

10. a. Mary drank 2 cups of tea at the party.
    b. Mom bought 1 quart of yogurt for the four children.
    c. Ron was quite thirsty and so he drank a whole pint of water.
    d. The large pitcher can hold 2 quarts of juice.

11. Answers vary.

Puzzle corner: 20 quarts

Milliliters and Liters, p. 113

1. - 4. Student activities. Answers will vary.

5.

a. 1 L = 1,000 ml 2 L = 2,000 ml 6 L = 6,000 ml	b. 1 L 200 ml = 1,200 ml 5 L 490 ml = 5,490 ml 4 L 230 ml = 4,230 ml	c. 7 L 70 ml = 7,070 ml 4 L 3 ml = 4,003 ml 9 L 409 ml = 9,409 ml

6. It contains 522 more ml of shampoo.

7. They contain 1 L 350 ml of water.

8. You can fill four 250 ml glasses or five 200 ml glasses from one liter.

9. There is 750 ml left in the pitcher.

Mixed Review Chapter 8, p. 115

1. a. Estimate: $150 + $130 = $280; Exact: $154 + $128 = $282.
   b. Estimate: $1,300 + $1,300 = $2,600; Exact: $1,298 + $1,298 = $2,596.
   c. Estimate: $1,300 − $800 = $500; Exact: $1,255 − $787 = $468.

2. Area = 90 × 6 = 540 ft^2

3. 20 × 9 = 180. Half of 180 is 90 so the area of one part is 90 ft^2.

4. a. 2,777, 2,778, 2,779   b. 6,059, 6,060, 6,061
   c. 7,149, 7,150, 7,151   d. 6,999, 7,000, 7,001

5.

a. 5 × 6 = 30 3 × 6 = 18 8 × 9 = 72 7 × 7 = 49	b. 6 × 7 = 42 4 × 7 = 28 5 × 12 = 60 8 × 12 = 96	c. 9 × 9 = 81 8 × 8 = 64 6 × 9 = 54 6 × 12 = 72

6.

a. 7 × 30 = 7 × 3 × 10 = 21 × 10 = 210	b. 5 × 60 = 5 × 6 × 10 = 30 × 10 = 300

7.

a. 8 × 70 = 560	b. 3 × 80 = 240	c. 50 × 4 = 200
d. 30 × 9 = 270	e. 20 × 6 = 120	f. 4 × 90 = 360

8. Total area = 560 m^2 Area of each part = 80 m^2

9. a. 4,946 Check: 4,945 + 2,316 = 7,262   b. 2,761 Check: 2,761 + 3,242 = 6,003

Review Chapter 8, p. 117

1. a. ─────────────
   b. ──────────

2. AB: _5_ cm _1_ mm
   BC: _7_ cm _2_ mm
   CA: _9_ cm _2_ mm
   perimeter: _21_ cm _5_ mm

   However, if you printed the lesson yourself, and didn't print at 100% but with "shrink to fit," "print to fit," or similar, the measurements will be smaller numbers than those given above. Please check the student's answers.
   For example, the student might get:

   AB: _4_ cm _7_ mm
   BC: _6_ cm _8_ mm
   CA: _8_ cm _6_ mm
   perimeter: _20_ cm _1_ mm

3. AB: _1 ½_ in   BC: _1_ in
   CD: _1 ½_ in   DA: _1_ in
   perimeter: _5_ in

   However, if you printed the lesson yourself, and didn't print at 100% but with "shrink to fit," "print to fit," or similar, the measurements will be smaller numbers than those given above. Please check the student's answers.

4. mm, cm, m, km

5. in, ft, yd, mi

Review Chapter 8, cont.

6. C, pt, qt, gal

7. pounds or kilograms

8. a. A butterfly's wings were 6 _cm_ wide.
   c. Jessica jogged 5 _km or mi_ yesterday.
   e. The distance from the city
     to the little town is 80 _km or mi._
   b. Sherry is 66 _in_ tall.
   d. The box was 60 _cm_ tall.
   f. The room was 4 _m_ wide.
   g. The eraser is 3 _cm_ long

9. a. 2 lb 12 oz   b. 2 lb 4 oz   c. 5 lb 12 oz

10. Answers will vary.

11. Answers will vary.

12. a. Mom bought 5 _kg or lb_ of apples.
    c. Dr. Smith weighs about 70 _kg_.
    e. The pan holds 2 _qt or L_ of water.
    b. Mary drank 350 _ml_ of juice.
    d. The banana weighed 3 _oz_.
    f. A cell phone weighs about 100 _g_.

Chapter 9: Division

Division as Making Groups, p. 122

1.

a. There are <u>15</u> carrots. Make groups of 5. How many groups? 3 How many 5's are there in <u>15</u>? 3	b. There are 20 berries. Make groups of 4. How many groups? 5 How many 4's are there in 20? 5	c. There are 9 apples. Make groups of 3. How many groups? 3 How many 3's are there in 9? 3
d. There are 10 fish. Make groups of 2. How many groups? 5 How many 2's are there in 10? 5	e. There are 12 daisies. Make groups of 6. How many groups? 2 How many 6's are there in 12? 2	f. There are 16 camels. Make groups of 4. How many groups? 4 How many 4's are there in 16? 4

2. a. $15 \div 5 = 3$. b. $20 \div 4 = 5$. c. $9 \div 3 = 3$. d. $10 \div 2 = 5$. e. $12 \div 6 = 2$. f. $16 \div 4 = 4$.

3. a. $10 \div 2 = 5$. b. $20 \div 4 = 5$. c. $18 \div 6 = 3$. d. $9 \div 3 = 3$. e. $15 \div 5 = 3$. f. $21 \div 3 = 7$.

4.

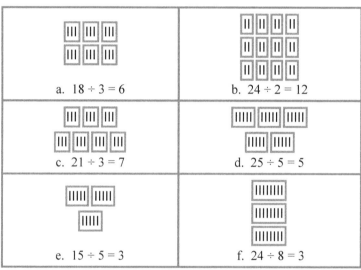

a. $18 \div 3 = 6$
b. $24 \div 2 = 12$
c. $21 \div 3 = 7$
d. $25 \div 5 = 5$
e. $15 \div 5 = 3$
f. $24 \div 8 = 3$

5. a. $20 \div 4 = 5$ b. $20 \div 2 = 10$ c. $30 \div 6 = 5$ d. $24 \div 3 = 8$
   e. $20 \div 5 = 4$ f. $21 \div 7 = 3$ g. $24 \div 6 = 4$ h. $20 \div 10 = 2$

6. a. $20 \div 5 = 4$ vans   b. $30 \div 5 = 6$ rows   c. $4 \times 20 = 80$ pins
   d. $28 \div 4 = 7$ bags   e. $16 \div 4 = 4$ big posters   f. $3 \times 7 = 21$ hours

7.

a.	b.	c.
$4 \div 2 = 2$	$20 \div 10 = 2$	$10 \div 5 = 2$
$6 \div 2 = 3$	$30 \div 10 = 3$	$15 \div 5 = 3$
$8 \div 2 = 4$	$40 \div 10 = 4$	$20 \div 5 = 4$
$10 \div 2 = 5$	$50 \div 10 = 5$	$25 \div 5 = 5$
$12 \div 2 = 6$	$60 \div 10 = 6$	$30 \div 5 = 6$
$14 \div 2 = 7$	$70 \div 10 = 7$	$35 \div 5 = 7$
$16 \div 2 = 8$	$80 \div 10 = 8$	$40 \div 5 = 8$
$18 \div 2 = 9$	$90 \div 10 = 9$	$45 \div 5 = 9$
$20 \div 2 = 10$	$100 \div 10 = 10$	$50 \div 5 = 10$

Division and Multiplication, p. 126

1. a. Two *groups of 6* is 12. $2 \times 6 = 12$

   12 divided into *groups of 6* is two groups. $12 \div 6 = 2$

   b. Five *groups of 2* is 10. $5 \times 2 = 10$

   10 divided into *groups of 2* is 5 groups. $10 \div 2 = 5$

   c. One *group of 4* is 4. $1 \times 4 = 4$

   4 divided into *groups of 4* is one group. $4 \div 4 = 1$

   d. 6 *groups of 3* is 18. $6 \times 3 = 18$

   18 divided into *groups of 3* is 6 groups. $18 \div 3 = 6$

   e. Five *groups of 1* is 5. $5 \times 1 = 5$

   5 divided into *groups of 1* is 5 groups. $5 \div 1 = 5$

   f. 7 *groups of 2* is 14. $7 \times 2 = 14$

   14 divided into *groups of 2* is 7 groups. $14 \div 2 = 7$

2. a. $2 \times 4 = 8$; $8 \div 4 = 2$. b. $6 \times 2 = 12$; $12 \div 2 = 6$.
   c. $4 \times 4 = 16$; $16 \div 4 = 4$. d. $2 \times 6 = 12$; $12 \div 6 = 2$.
   e. $5 \times 4 = 20$; $20 \div 4 = 5$. f. $2 \times 7 = 14$; $14 \div 7 = 2$.
   g. $3 \times 6 = 18$; $18 \div 6 = 3$. h. $4 \times 2 = 8$; $8 \div 2 = 4$. i. $1 \times 5 = 5$; $5 \div 5 = 1$.

3.

   a. $3 \times 5 = 15$
   $15 \div 5 = 3$

   b. $3 \times 8 = 24$
   $24 \div 8 = 3$

   c. $6 \times 5 = 30$
   $30 \div 5 = 6$

   d. $3 \times 9 = 27$
   $27 \div 9 = 3$

   e. $2 \times 16 = 32$
   $32 \div 16 = 2$

   f. $8 \times 2 = 16$
   $16 \div 2 = 8$

   g. $1 \times 8 = 8$
   $8 \div 8 = 1$

   h. $2 \times 9 = 18$
   $18 \div 9 = 2$

   i. $4 \times 5 = 20$
   $20 \div 5 = 4$

4. a. 14; $14 \div 2 = 7$ b. 24; $24 \div 2 = 12$ c. 40; $40 \div 5 = 8$
   d. 42; $42 \div 7 = 6$ e. 49; $49 \div 7 = 7$ f. 33; $33 \div 3 = 11$
   g. 72; $72 \div 8 = 9$ h. 5; $5 \div 5 = 1$ i. 63; $63 \div 9 = 7$

5. a. 7; $7 \times 2 = 14$ b. 9; $9 \times 2 = 18$ c. 3; $3 \times 7 = 21$
   d. 9; $9 \times 6 = 54$ e. 6; $6 \times 4 = 24$ f. 10; $10 \times 3 = 30$
   g. 8; $8 \times 4 = 32$ h. 8; $8 \times 7 = 56$ i. 11; $11 \times 5 = 55$

6. a. 6, 8, 10, 4.   b. 3, 7, 7, 6.   c. 4, 5, 10, 7.   d. 8, 7, 9, 10.

Puzzle corner. a. 10   b. 8   c. 50   d. 2   e. 1   f. 5

Division and Multiplication Facts, p. 130

1. a. $4 \times 6 = 24$
      $6 \times 4 = 24$
      $24 \div 4 = 6$
      $24 \div 6 = 4$

   b. $3 \times 5 = 15$
      $5 \times 3 = 15$
      $15 \div 5 = 3$
      $15 \div 3 = 5$

   c. $7 \times 4 = 28$
      $4 \times 7 = 28$
      $28 \div 7 = 4$
      $28 \div 4 = 7$

   d. $5 \times 4 = 20$
      $4 \times 5 = 20$
      $20 \div 5 = 4$
      $20 \div 4 = 5$

2. a. $9 \times 2 = 18$
      $2 \times 9 = 18$
      $18 \div 2 = 9$
      $18 \div 0 = 2$

   b. $3 \times 9 = 27$
      $9 \times 3 = 27$
      $27 \div 3 = 9$
      $27 \div 9 = 3$

   c. $1 \times 5 = 5$
      $5 \times 1 = 5$
      $5 \div 5 = 1$
      $5 \div 1 = 5$

   d. $2 \times 6 = 12$
      $6 \times 2 = 12$
      $12 \div 2 = 6$
      $12 \div 6 = 2$

3.

a. $12 \div 2 = 6$ $6 \times 2 = 12$	b. $22 \div 2 = 11$ $11 \times 2 = 22$	c. $16 \div 2 = 8$ $8 \times 2 = 16$
d. $24 \div 3 = 8$ $8 \times 3 = 24$	e. $32 \div 4 = 8$ $8 \times 4 = 32$	f. $25 \div 5 = 5$ $5 \times 5 = 25$

4. a. 9, 8, 12   b. 5, 6, 7   c. 10, 4, 9   d. 9, 11, 12

5.

a. $14 \div 2 = 7$ $7 \times 2 = 14$	b. $30 \div 5 = 6$ $6 \times 5 = 30$	c. $28 \div 7 = 4$ $4 \times 7 = 28$
d. $20 \div 10 = 2$ $2 \times 10 = 20$	e. $35 \div 7 = 5$ $5 \times 7 = 35$	f. $56 \div 7 = 8$ $8 \times 7 = 56$

6. In each problem, the other matching multiplication and division is also correct.
   a. $5 \times 10 = 50$; $50 \div 5 = 10$ b. $4 \times 5 = 20$; $20 \div 4 = 5$
   c. $3 \times 20 = 60$; $60 \div 20 = 3$ d. $3 \times 10 = 30$; $30 \div 3 = 10$

7. a. 8, 10, 6, 7, 4.   b. 8, 4, 11, 12, 9.   c. 45, 27, 18, 81, 99.   d. 8, 12, 9, 10, 4.

Dividing Evenly into Groups, p. 133

1. a. $12 \div 2 = 6$ b. $6 \div 2 = 3$ c. $10 \div 2 = 5$

2. a. $12 \div 3 = 4$ b. $6 \div 3 = 2$ c. $24 \div 3 = 8$

3. a. $8 \div 4 = 2$ b. $12 \div 4 = 3$ c. $24 \div 4 = 6$

4. a. $8 \div 2 = 4$ b. $10 \div 5 = 2$ c. $15 \div 1 = 15$ d. $20 \div 4 = 5$
   e. $21 \div 3 = 7$ f. $21 \div 1 = 21$ g. $30 \div 10 = 3$ h. $14 \div 2 = 7$

5. a. 5, 2, 8   b. 4, 10, 5   c. 4, 6, 8   d. 6, 8, 8   e. 11, 12, 9   f. 10, 8, 5

6. a. 4   b. 7   c. 7   d. 60   e. 28   f. 9   g. 12   h. 121

Dividing Evenly into Groups, cont.

7.

a. $18 \div 3 = 6$ They each got six marbles.	b. $4 \times 7 = 28$ There was a total of 28 marbles.
c. $24 \div 6 = 4$ The pieces were 4 inches long.	d. $24 \div 3 = 8$ Each girl got 8 hairpins.

8. a. and b. Answers will vary. Please check the student's work.

9. a. Division table of six	b. Division table of seven	c. Division table of eight
$6 \div 6 = 1$	$7 \div 7 = 1$	$8 \div 8 = 1$
$12 \div 6 = 2$	$14 \div 7 = 2$	$16 \div 8 = 2$
$18 \div 6 = 3$	$21 \div 7 = 3$	$24 \div 8 = 3$
$24 \div 6 = 4$	$28 \div 7 = 4$	$32 \div 8 = 4$
$30 \div 6 = 5$	$35 \div 7 = 5$	$40 \div 8 = 5$
$36 \div 6 = 6$	$42 \div 7 = 6$	$48 \div 8 = 6$
$42 \div 6 = 7$	$49 \div 7 = 7$	$56 \div 8 = 7$
$48 \div 6 = 8$	$56 \div 7 = 8$	$64 \div 8 = 8$
$54 \div 6 = 9$	$63 \div 7 = 9$	$72 \div 8 = 9$
$60 \div 6 = 10$	$70 \div 7 = 10$	$80 \div 8 = 10$
$66 \div 6 = 11$	$77 \div 7 = 11$	$88 \div 8 = 11$
$72 \div 6 = 12$	$84 \div 7 = 12$	$96 \div 8 = 12$
The patterns are like the multiplication tables "backwards."		

Division Word Problems, p. 137

1.

a. $90 \div 10 = 9$ Nine pages are full of stamps.	b. $12 \times 8 = 96$ She has 96 stamps.
c. $4 \times 11 = 44$ There would be 44 children.	d. $12 \div 4 = 3$ You would need three taxis.

2.

a. $10 \times 5 = 50$ There are 50 eggs.	b. $10 \div 5 = 2$ He used 2 bags.
c. $3 \times 5 = 15$ She can fit 15 bottles of juice.	d. $18 \div 3 = 6$ She will need six bags.
e. $36 \div 3 = 12$ Each one got 12 cherries.	f. $25 \div 5 = 5$ Five students were in each group.
g. $7 \times 5 = 35$ There were 35 people.	h. $27 \div 3 = 9$ Each part was nine inches long.
i. $20 \div 2 = 10$ She filled ten containers.	j. $9 \div 3 = 3$ Each part was three miles long.
k. $24 \div 8 = 3$ She can make three omelets.	l. $60 \div 12 = 5$ You will need five boxes.

3. a. There were 78 chairs. $7 \times 10 + 8 = 78$
   b. Each got 9 cherries. $14 + 13 = 27;\ 27 \div 3 = 9$
   c. She used 11 containers. $4 + 6 + 7 + 5 = 22;\ 22 \div 2 = 11$
   d. There are 55 crayons. $5 \times 12 - 5 = 55$

Division Word Problems, cont.

4.

a. $4 \times 9 = 36$	b. $9 \times 6 = 54$	c. $6 \times 7 = 42$
$9 \times 4 = 36$	$6 \times 9 = 54$	$7 \times 6 = 42$
$36 \div 9 = 4$	$54 \div 6 = 9$	$42 \div 7 = 6$
$36 \div 4 = 9$	$54 \div 9 = 6$	$42 \div 6 = 7$

Puzzle corner: There are many possible answers. These are just some examples.

20	÷	4	= 5
÷		÷	
5	÷	1	= 5
=		=	
4		4	

54	÷	9	= 6
÷		÷	
6	÷	3	= 2
=		=	
9		3	

Zero in Division, p. 140

1.

| a. $4 \div 1 = 4$
 ~~$4 \div 0 =$~~ ___ | b. $14 \div 14 = 1$
 ~~$0 \div 0 =$~~ ___ | c. $1 \div 1 = 1$
 ~~$7 \div 0 =$~~ ___ | d. $0 \div 5 = 0$
 $5 \div 5 = 1$ |
| e. $0 \div 1 = 0$
 $0 \div 4 = 0$ | f. $0 \div 14 = 0$
 ~~$14 \div 0 =$~~ ___ | g. $0 \div 3 = 0$
 $0 \div 1 = 0$ | h. $10 \div 10 = 1$
 $1 \div 1 = 1$ |

2.

a. $6 \times 1 = 6$ $6 \div 1 = 6$	b. $0 \times 8 = 0$ $0 \div 8 = 0$	c. $5 \times 7 = 35$ $35 \div 7 = 5$
d. $10 \times 11 = 110$ $110 \div 11 = 10$	e. $1 \times 1 = 1$ $1 \div 1 = 1$	f. $1 \times 8 = 8$ $8 \div 8 = 1$
g. $0 \times 0 = 0$ not possible	h. $5 \times 9 = 45$ $45 \div 9 = 5$	i. $9 \times 0 = 0$ not possible

3. a. $30 \div 6 = 5$ kg   b. $7 \times 6 = 42$ passengers.
   c. $3 \times 12 = 36$ eggs   d. $56 \div 7 = 8$ minivans.
   e. $5 \times 4 = 20$ legs   f. $4 \div 4 = 1$ cup of milk in each glass.

4.

| a. $9 \div 1 = 9$
 ~~$9 \div 0 =$~~ | b. $0 \div 20 = 0$
 ~~$20 \div 0 =$~~ | c. $11 \div 1 = 11$
 ~~$8 \div 0 =$~~ | d. ~~$0 \div 0$~~
 $0 \div 10 = 0$ |

5. a. How many cars did each boy get? $18 + 7 + 11 = 36$. $36 \div 3 = 12$. Each boy got 12 cars.
   b. How much did she pay them in total? $5 \times 10 + 15 = \$65$. She paid them $65 in total.
   c. How many pages does Alice need to read? $320 + 129 + 120 + 235 = 804$. She needs to read 804 pages.
   d. How many days will it take him to read the books? $32 + 40 = 72$. $72 \div 12 = 6$.
      It will take him six days to read the books.
   e. How many pieces will Kelly get? $80 \div 20 = 4$ and $40 \div 20 = 2$. $4 + 2 = 6$. She gets 6 pieces.
   f. How many cars did the child have? $7 \times 6 + 3 = 45$. He had 45 cars.

Puzzle corner: a. no solutions b. Any number is a solution (there are an infinite number of solutions) c. & d. No solutions.

When Division is Not Exact, p. 143

1. c. $8 \div 5 = \underline{1}$, remainder $\underline{3}$.   d. remainder $\underline{3}$
   e. $\underline{15}$ rams divided among 6 people gives $\underline{2}$ rams to each and $\underline{3}$ rams that cannot be divided. $15 \div 6 = 2$, remainder 3.
   f. $\underline{9}$ camels divided between 2 people gives $\underline{4}$ camels to each person, and $\underline{1}$ camel left over. $9 \div 2 = 4$, remainder 1.

2. a. $20 \div 3 = 6$ R2   b. $21 \div 4 = 5$ R1   c. $21 \div 6 = 3$ R3   d. $24 \div 5 = 4$ R4
   e. $24 \div 7 = 3$ R3   f. $20 \div 9 = 2$ R2   g. $16 \div 3 = 5$ R1   h. $16 \div 5 = 3$ R1

3. a. 3 R1; 0 R1   b. 0 R3; 5 R1   c. 3 R3; 1 R1

4. a. 2 R3, 2 R4   b. 0 R5, 3 R1.   c. 7 R5, 8 R2
   d. 6 R1, 2 R3   e. 6 R6, 8 R3   f. 8 R1, 4 R3

5.

The pattern is: the remainders go cyclically from 0 to one less than the divisor.

The quotient (answers) remains the same for each different remainder, then increases by 1.

a.	b.	c.
$21 \div 2 = 10$, R1	$21 \div 3 = 7$, R0	$21 \div 4 = 5$, R1
$22 \div 2 = 11$, R0	$22 \div 3 = 7$, R1	$22 \div 4 = 5$, R2
$23 \div 2 = 11$, R1	$23 \div 3 = 7$, R2	$23 \div 4 = 5$, R3
$24 \div 2 = 12$, R0	$24 \div 3 = 8$, R0	$24 \div 4 = 6$, R0
$25 \div 2 = 12$, R1	$25 \div 3 = 8$, R1	$25 \div 4 = 6$, R1
$26 \div 2 = 13$, R0	$26 \div 3 = 8$, R2	$26 \div 4 = 6$, R2
$27 \div 2 = 13$, R1	$27 \div 3 = 9$, R0	$27 \div 4 = 6$, R3
$28 \div 2 = 14$, R0	$28 \div 3 = 9$, R1	$28 \div 4 = 7$, R0
$29 \div 2 = 14$, R1	$29 \div 3 = 9$, R2	$29 \div 4 = 7$, R1
$30 \div 2 = 15$, R0	$30 \div 3 = 10$, R0	$30 \div 4 = 7$, R2

More Practice with the Remainder, p. 146

1. a. 8   b. 8   c. 4   d. 8   e. 7   f. 7   g. 9   h. 6

2. a. 7 R1, $3 \overline{)22}$, -21, 1
   b. 4 R1, $5 \overline{)21}$, -20, 1
   c. 5 R2, $3 \overline{)17}$, -15, 2
   d. 3 R5, $8 \overline{)29}$, -24, 5
   e. 3 R5, $7 \overline{)26}$, -21, 5
   f. 8 R4, $6 \overline{)52}$, -48, 4
   g. 3 R8, $9 \overline{)35}$, -27, 8
   h. 8 R3, $4 \overline{)35}$, -32, 3

3. a. 9 R1, $3 \overline{)28}$, -27, 1
   b. 9 R3, $5 \overline{)48}$, -45, 3
   c. 9 R4, $6 \overline{)58}$, -54, 4
   d. 8 R7, $8 \overline{)71}$, -64, 7
   e. 8 R5, $7 \overline{)61}$, -56, 5
   f. 7 R4, $6 \overline{)46}$, -42, 4
   g. 8 R3, $9 \overline{)75}$, -72, 3
   h. 7 R5, $7 \overline{)54}$, -49, 5

4. a. $29 \div 8 = 3$, R5. So each guest got 3 rolls, and 5 were left over.
   b. $5 \times 7 = 35$ plants.
   c. $11 \times 6 + 3 = 69$ pencils. OR $12 \times 6 - 3 = 69$ pencils. Notice there are 11 students plus Jim.
   d. $3 \times 12 = 36$ rolls
   e. Of three, yes. $12 \div 3 = 4$. Of four, yes. $12 \div 4 = 3$. Of five, no. $12 \div 5 = 2$ R2. Of six, yes. $12 \div 6 = 2$.
   f. He can make rows of 18, rows of 12, rows of 9, rows of 6, rows of 4, or rows of 3 plants. He can also make rows of 2 plants or rows of 1 plant, though it is less likely he would choose these options.
   g. $40 \div 15 = 2$ R10. So there are 2 lollipops for each child and 10 are left over.
   h. There are four full cartons and one carton with only 8 eggs in it. $4 \times 12 + 8 = 56$ or $56 \div 12 = 4$ R8.

Mixed Review Chapter 9, p. 148

1. Area 8 in². Perimeter 12 in.

2. a. She has 40 minutes left.   b. It took him 28 minutes.   c. It is 13 minutes till the class ends.

3.

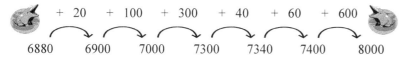

4. a. VIII, XII   b. XIX, XXIV   c. XL, XLIV   d. XC, LXXVI

5.

a.	b.	c.	d.
416 ≈ 400	529 ≈ 500	670 ≈ 700	254 ≈ 300
837 ≈ 800	960 ≈ 1000	557 ≈ 600	147 ≈ 100

6. a. Estimate: $250 − $70 = $180   Exact: $245 − $68 = $177
   b. The total cost was $354. $400 − $177 − $177 = $46 change.

7. Check the student's answers.
   a. _____
   b. _____
   c. _____
   d. _____

8. a. Ann's living room is 20 ft wide.
   c. It is about 2 mi to the bookstore.
   b. The refrigerator is 28 in wide.
   d. The doctor is 6 ft tall.

9. a. The fly was 12 mm long.
   c. Mark rode his bicycle 12 km to go home.
   b. The room measures about 3 m.
   d. The teddy bear was 25 cm tall.

Review Chapter 9, p. 150

1. 2 × 6 = 12   12 ÷ 6 = 2   b. 3 × 5 = 15   15 ÷ 5 = 3

2.

a.	b.	c.	d.
36 ÷ 6 = 6 3 ÷ 3 = 1 36 ÷ 3 = 12 4 ÷ 1 = 4	44 ÷ 11 = 4 60 ÷ 6 = 10 25 ÷ 5 = 5 54 ÷ 9 = 6	56 ÷ 7 = 8 72 ÷ 9 = 8 99 ÷ 9 = 11 100 ÷ 10 = 10	0 ÷ 9 = 0 16 ÷ 16 = 1 12 ÷ 1 = 12 12 ÷ 2 = 6

3.

a.	b.	c.
7 × 6 = 42 6 × 7 = 42 42 ÷ 6 = 7 42 ÷ 7 = 6	8 × 1 = 8 1 × 8 = 8 8 ÷ 1 = 8 8 ÷ 8 = 1	7 × 7 = 49 7 × 7 = 49 49 ÷ 7 = 7 49 ÷ 7 = 7

4. a. 9   b. 20   c. 9   d. 9   e. 12   f. 7   g. 6   h. 64

5.

a. 6 × 0 = 0 0 ÷ 6 = 0	b. 1 × 9 = 9 9 ÷ 1 = 9	c. 0 × 0 = 0 ~~0 ÷ 0~~

6.

a. 11 ÷ 2 = 5 R1	b. 41 ÷ 8 = 5 R1	c. 16 ÷ 5 = 3 R1
d. 56 ÷ 10 = 5 R6	e. 26 ÷ 4 = 6 R2	f. 22 ÷ 9 = 2 R4

7. a. 6 × 8 = 48 She has 48 crayons.
   b. 24 ÷ 6 = 4 There were four groups of six children.
   c. 48 ÷ 6 = 8 She had eight bags of cookies.
   d. 94 ÷ 10 = 9 R 4. (Or, 9 × 10 + 4 = 94), so 9 pages are full of stamps. The tenth page has 4 stamps on it.

Puzzle corner:

80	÷	8	=	10
÷		÷		
10	÷	2	=	5
=		=		
8		4		

54	÷	9	=	6
÷		÷		
6	÷	3	=	2
=		=		
9		3		

Chapter 10: Fractions

Understanding Fractions. p. 155

2. a. 1/3   b. 1/5   c. 3/4   d. 2/8   e. 2/5   f. 4/7   g. 5/10   h. 3/12   i. 7/12   j. 3/6   k. 4/18   l. 7/8

3. a. 2/5; two fifths   b. 4/5; four fifths   c. 3/4; three fourths
   d. 5/8; five eighths   e. 5/6; five sixths   f. 8/8; eight eighths

5.

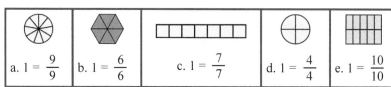

6.

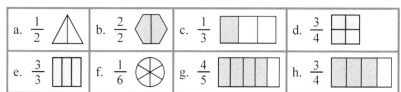

7.

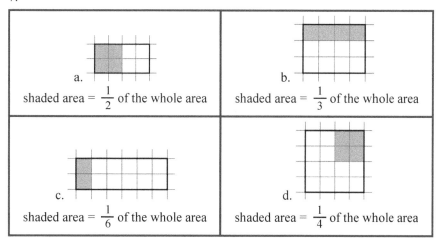

Understanding Fractions, cont.

7. continued.

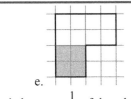
e.
shaded area = $\frac{1}{3}$ of the whole area

f.
shaded area = $\frac{1}{5}$ of the whole area

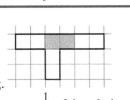
g.
shaded area = $\frac{1}{4}$ of the whole area

h.
shaded area = $\frac{1}{4}$ of the whole area

Fractions on a Number Line, p. 159

1. a. The number line from 0 to 1 is divided into _5_ parts. The arrow marks the fraction 4/5.

   b. The number line from 0 to 1 is divided into _6_ parts. The arrow marks the fraction 2/6.

2. a. Number line from 0 to 1 divided into fourths: $\frac{0}{4}, \frac{1}{4}, \frac{2}{4}, \frac{3}{4}, \frac{4}{4}$

   b. Number line from 0 to 1 divided into sevenths: $\frac{0}{7}, \frac{1}{7}, \frac{2}{7}, \frac{3}{7}, \frac{4}{7}, \frac{5}{7}, \frac{6}{7}, \frac{7}{7}$

3. a. 1/3   b. 2/9   c. 6/10   d. 4/6

4.

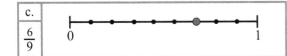

5.

Fractions on a Number Line, cont.

g. 4/5 — [number line 0 to 1, dot at 4/5]

h. 1/6 — [number line 0 to 1, dot at 1/6]

i. 5/6 — [number line 0 to 1, dot at 5/6]

j. 3/8 — [number line 0 to 1, dot at 3/8]

6. Number line 0 to 3 showing 0/5, 1/5, 2/5, 3/5, 4/5, 5/5, 6/5, 7/5, 8/5, 9/5, 10/5, 11/5, 12/5, 13/5, 14/5, 15/5

7. a. Number line 0 to 3 showing 3/6, 7/6, 11/6, 13/6, 18/6

18/6 is the whole number 3.

b. Number line 0 to 3 showing 5/8, 12/8, 16/8, 17/8, 21/8

16/8 is the whole number 2.

8. a. 1 = 6/6   b. 2 = 16/8   c. 3 = 12/4   d. 2 = 12/6   e. 3 = 9/3   f. 1 = 7/7   g. 4 = 20/5   h. 4 = 32/8

Can you find a shortcut so that you do not actually have to *count* all those pie pieces?
Multiply the number of pies by the number of pieces in each.

9.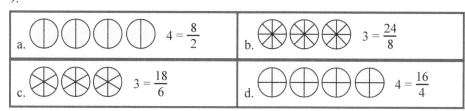

a. 4 = 8/2   b. 3 = 24/8   c. 3 = 18/6   d. 4 = 16/4

10. a. 1   b. 3   c. 4   d. 10   e. 5   f. 1   g. 4   h. 6

Puzzle corner. a. 5 = 30/6   b. 7 = 35/5   c. 3 = 21/7   d. 6 = 60/10   e. 9 = 45/5

Mixed Numbers, p. 163

1. a. 1 2/5   b. 2 4/6   c. 1 2/3   d. 3 5/10   e. 4 1/3   f. 5 2/9

2.

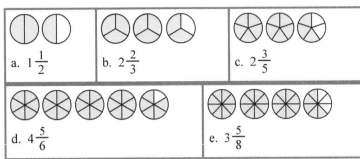

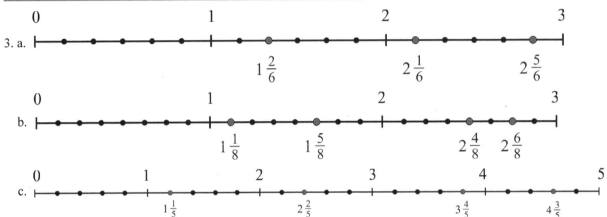

4. a. 1 2/6   b. 2 1/6   c. 2 5/6   d. 1 3/8   e. 2 5/8   f. 2 3/5   g. 1 4/5   h. 4 1/5

5.

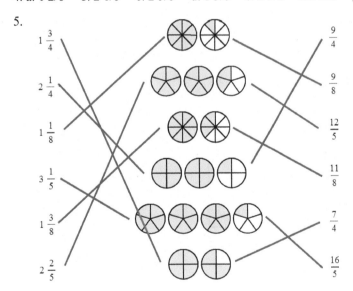

6. a. 3 1/2 = 7/2   b. 1 2/10 = 12/10   c. 2 3/4 = 11/4   d. 3 0/3 = 9/3   e. 2 1/6 = 13/6   f. 2 5/10 = 25/10.

7. a. 1 2/4 L   b. 1 3/4 L   c. 2 1/4 L

8. a. 1 3/4 inches   b. 3 1/4 inches   c. 4 3/4 inches

9. a. ───────────────────────────────
   b. ──────────────────────────────────────

Equivalent Fractions, p. 167

1. a. $\frac{1}{4} = \frac{2}{8}$ b. $\frac{2}{3} = \frac{6}{9}$ c. $\frac{4}{8} = \frac{1}{2}$ d. $\frac{4}{6} = \frac{2}{3}$ e. $\frac{2}{3} = \frac{8}{12}$ f. $\frac{5}{6} = \frac{10}{12}$

2. a. $\frac{1}{3} = \frac{3}{9}$ b. $\frac{6}{8} = \frac{3}{4}$

3.
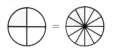 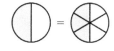 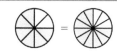 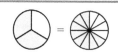

a. $\frac{1}{4} = \frac{3}{12}$ b. $\frac{1}{2} = \frac{3}{6}$ c. $\frac{6}{8} = \frac{9}{12}$ d. $\frac{2}{3} = \frac{8}{12}$

e. $\frac{1}{3} = \frac{2}{6}$ f. $\frac{8}{12} = \frac{4}{6}$

4.
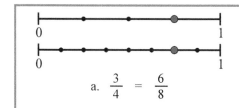 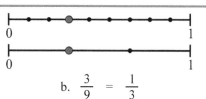

a. $\frac{3}{4} = \frac{6}{8}$ b. $\frac{3}{9} = \frac{1}{3}$

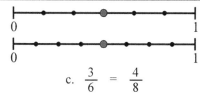 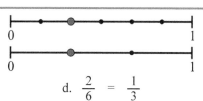

c. $\frac{3}{6} = \frac{4}{8}$ d. $\frac{2}{6} = \frac{1}{3}$

5.
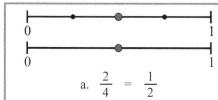 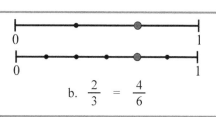

a. $\frac{2}{4} = \frac{1}{2}$ b. $\frac{2}{3} = \frac{4}{6}$

Equivalent Fractions, cont.

6.

a. | | $\frac{1}{3}$ | b. | | $\frac{1}{4}$
| | $\frac{2}{6}$ | | | $\frac{2}{8}$
| | $\frac{3}{9}$ | | | $\frac{3}{12}$
| | $\frac{4}{12}$ | | | $\frac{4}{16}$

7. Both ideas work, and give everybody a fair share. Cassy's idea gives everyone 1/4 of the bar. Hannah's idea gives everyone 3/12 of the bar. But 1/4 and 3/12 are equivalent fractions, so they signify the same amount of the chocolate bar.

8. Answers vary. Students may use pie models, rectangular models, number lines, squares, or other shapes to show this.

$\frac{1}{2} = \frac{4}{8}$

9. a. (pie model)   b. $\frac{1}{2} = \frac{3}{6}$

10. Yes, they are. They are "the same amount of pie." Also, both are equal to 1, so are also equal to each other.

Puzzle corner. The line 3 1/2 inches long is longer. It is 1/4-inch longer than the other line.

Comparing Fractions 1, p. 170

1.

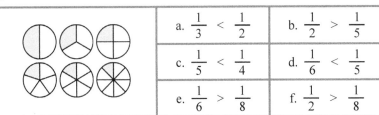

a. $\frac{1}{3} < \frac{1}{2}$		b. $\frac{1}{2} > \frac{1}{5}$
c. $\frac{1}{5} < \frac{1}{4}$		d. $\frac{1}{6} < \frac{1}{5}$
e. $\frac{1}{6} > \frac{1}{8}$		f. $\frac{1}{2} > \frac{1}{8}$

2. 1/10 is the smallest fraction.
   1/2 is the greatest fraction.

3. Since 1/3 is further from zero than 1/4, it is the bigger fraction.

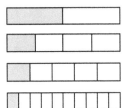

4. One eighth (1/8) is the bigger fraction. If you divide a whole into 8 pieces, each piece is bigger than if you divide a whole into 9 pieces.

5. $\frac{1}{9} < \frac{1}{6} < \frac{1}{5} < \frac{1}{3}$

Comparing Fractions 1, cont.

6.

a. $\frac{2}{3} > \frac{1}{3}$	b. $\frac{1}{5} < \frac{4}{5}$	c. $\frac{1}{6} < \frac{3}{6}$
d. $\frac{1}{9} < \frac{5}{9}$	e. $\frac{5}{12} > \frac{3}{12}$	f. $\frac{6}{10} < \frac{7}{10}$

7. You can just compare how many parts there are. For example, 5/8 has more eighths than 3/8, so it is the bigger fraction. In other words, just compare the numerators.

8.

a. $\frac{2}{6} < \frac{2}{3}$	b. $\frac{2}{5} > \frac{2}{8}$	c. $\frac{3}{6} < \frac{3}{4}$
d. $\frac{5}{6} > \frac{5}{8}$	e. $\frac{2}{2} > \frac{2}{3}$	f. $\frac{4}{8} < \frac{4}{5}$
g. $\frac{8}{10} < \frac{8}{9}$	h. $\frac{3}{8} < \frac{3}{6}$	i. $\frac{7}{12} < \frac{7}{9}$

9. You can just check what kind of pieces the two fractions have, and choose the fraction that has bigger pieces. For example, 5/7 has sevenths, and sevenths are bigger pieces than eighths, so 5/7 is more than 5/8. In other words, just compare the denominators, and the fraction with the smaller denominator is the greater fraction.

10.

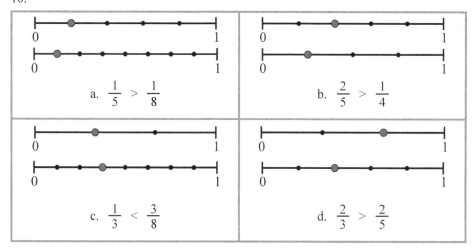

Comparing Fractions 1, cont.

11.

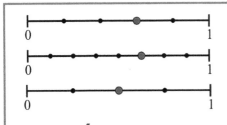
a. $\frac{5}{8}$ is the biggest.

b. $\frac{7}{8}$ is the biggest.

12. a. $\frac{5}{7} > \frac{4}{7}$ b. $\frac{7}{5} > \frac{7}{10}$ c. $\frac{9}{12} > \frac{5}{12}$ d. $\frac{10}{20} < \frac{10}{12}$

e. $\frac{3}{7} < \frac{3}{4}$ f. $\frac{7}{11} < \frac{10}{11}$ g. $\frac{5}{8} < \frac{5}{6}$ h. $\frac{5}{8} < \frac{9}{8}$

Comparing Fractions 2, p. 173

1. a. No. The bigger pitcher has more.
   b. It is hard to tell who got more pie to eat. You cannot really tell. Notice that the wholes are not the same size.

2. a. (cross out) b. 2/9 < 2/6 c. cross out
   d. 3/10 > 1/4 e. (cross out) f. 2/7 > 2/9

3. No, it will not be fair. One-third of the large pizza is a bigger piece to eat than one-third of the small pizza.

4. One-twelfth of the bigger bar is more to eat than 2/12 of the smaller. However, that does not prove that 1/12 > 2/12 because we were not using wholes of the same size.

5. A piece from the red ribbon. That is because 1/4 > 1/5.

6. The big can has more paint in it.

7. a. > b. > c. > d. < e. < f. < g. < h. <

8. Margaret's number lines are not of the same length so she is using them incorrectly. When we make the two number lines to have the same length from 0 to 1, we see that 1/5 < 1/4.

   You can also use pie pictures: ⊘ < ⊕

Mixed Review Chapter 10, p. 175

1.

a.	b.	c.	d.
56 ÷ 7 = 8	48 ÷ 6 = 8	54 ÷ 9 = 6	48 ÷ 8 = 6
49 ÷ 7 = 7	72 ÷ 6 = 12	81 ÷ 9 = 9	72 ÷ 8 = 9
28 ÷ 7 = 4	54 ÷ 6 = 9	36 ÷ 9 = 4	32 ÷ 8 = 4

2.

a. 7 × 6 = 42	b. 3 × 0 = 0	c. 9 × 8 = 72
42 ÷ 7 = 6	0 ÷ 3 = 0	72 ÷ 9 = 8
42 ÷ 6 = 7	~~3 ÷ 0~~ (not possible)	72 ÷ 8 = 9

3.

a.	b.	c.
16 ÷ 5 = 3 R1	21 ÷ 4 = 5 R1	19 ÷ 6 = 3 R1
12 ÷ 5 = 2 R2	27 ÷ 4 = 6 R3	31 ÷ 6 = 5 R1

4. Kathy needs to read nine pages of the book each day.

5. Each child will get three apples and there will be two apples left over.

6. 2 × 3 + 4 × 6 = 30   OR   6 × 3 + 4 × 3 = 30   OR   6 × 6 − 2 × 3 = 30

7. Estimate: 140 ft + 80 ft + 140 ft + 80 ft = 440 ft.
   The real perimeter: 438 ft

8. a. 1/4   b. 2/7   c. 3/8   d. 6/7

9. Since the two fractions have the same amount of pieces, look at the size of the pieces. Eighths are larger pieces than tenths, so 7/8 is greater than 7/10.

10. a. 5,637   b. 6,121   c. 9,696   d. 4,010

Fractions Review, p. 177

1.

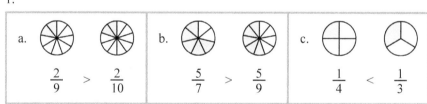

2.

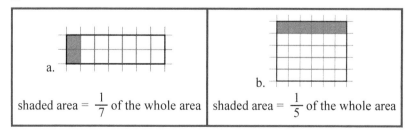

3. a. 1/5   b. 2/10   c. 8/9   d. 6/8

Fractions Review, cont.

4. a.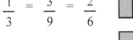
   b.

5.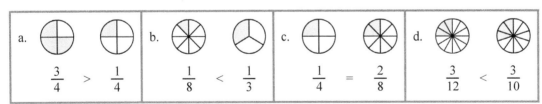

6. a. 1 1/3   b. 2 2/3   c. 2   d. 3

7. a. 1 = 7/7   b. 2 = 6/3   c. 4 = 20/5

8. a. 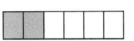   b.

9.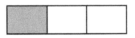

   $\frac{1}{3} = \frac{3}{9} = \frac{2}{6}$

10.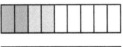

    $\frac{2}{9} < \frac{1}{3} < \frac{4}{9} < \frac{3}{6}$

11.

| a. $\frac{3}{4} > \frac{1}{4}$ | b. $\frac{1}{8} < \frac{1}{3}$ | c. $\frac{1}{4} = \frac{2}{8}$ | d. $\frac{3}{12} < \frac{3}{10}$ |

12. a. <   b. >   c. <   d. =

13. Since the two fractions have the same kinds of pieces (ninths), look at the amount of the pieces. Eight pieces is more than five pieces, so 8/9 is greater than 5/9.

Fractions Review, cont.

14.

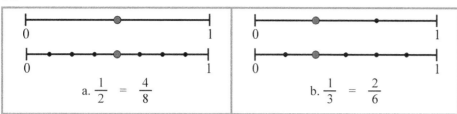

15.

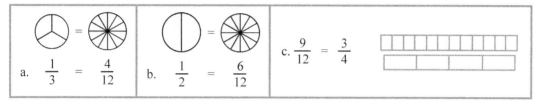

16. No. She is not correct. The two wholes are not the same size. If the wholes are made the same size, we can easily see that 2/8 < 2/4.

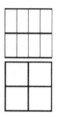

Test Answer Keys

Math Mammoth Grade 3 Tests Answer Key

Chapter 1 Test

Grading

My suggestion for grading the chapter 1 test is below. The total is 39 points. Divide the student's score by the total of 39 to get a decimal number, and change that decimal to percent to get the student's percentage score.

Question	Max. points	Student score
1	6 points	
2	3 points	
3	6 points	
4	3 points	
5	4 points	
6	4 points	

Question	Max. points	Student score
7	4 points	
8	2 points	
9	3 points	
10	2 points	
11	2 points	
Total	39 points	

1. a. 270; 203   b. 93;129   c. 47; 871

2. a. 5   b. 287   c. 8

3. a. 4   b. 66   c. 78   d. 144   e. 29   f. 98

4. **He has $105 left.** His purchases were a total of $145. And, $250 − $145 = $105.

5. a. **247.** To check it, add 247 + 157 = 404.   b. **326.** To check it, add 326 + 397 = 723

6. a. 710   b. 600   c. 820   d. 460

7. a. 27   b. 43   c. 310   d. 320

8. 159 days

9. Jason has 4 × 80 − 28 = 320 − 28 = **292 trading cards**.

10.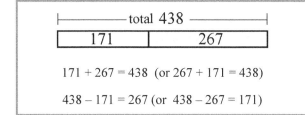

    171 + 267 = 438 (or 267 + 171 = 438)

    438 − 171 = 267 (or 438 − 267 = 171)

11. 295

101

Chapter 2 Test

Grading

My suggestion for grading the chapter 2 test is below. The total is 26 points. Divide the student's score by the total of 26 to get a decimal number, and change that decimal to percent to get the student's percentage score.

Question	Max. points	Student score
1	12 points	
2	4 points	

Question	Max. points	Student score
3	6 points	
4	4 points	
Total	26 points	

1. a. 6, 5, 0   b. 10, 30, 12   c. 40, 120, 400   d. 9, 0, 11

2. Answers will vary. Check students' answers. For example:

   a.   b.

3. a. There are 3 × 12 = **36 apples in three baskets**.
   b. 4 × $2 + 2 × $8 = $24
   c. **You can make five groups.** 5 × 4 = 20

4. a. 20   b. 22   c. 0   d. 11

Chapter 3 Test

Grading

My suggestion for grading the chapter 3 test is below. The total is 53 points. Divide the student's score by the total of 53 to get a decimal number, and change that decimal to percent to get the student's percentage score.

Question	Max. points	Student score
1	17 points	
2	4 points	

Question	Max. points	Student score
3	8 points	
4	24 points	
Total	53 points	

1.

×	0	1	2	3	4	5	6	7	8	9	10	11	12
0	0	0	0	0	0	0	0	0	0	0	0	0	0
1	0	1	2	3	4	5	6	7	8	9	10	11	12
2	0	2	4	6	8	10	12	14	16	18	20	22	24
3	0	3	6	9	12	15	18	21	24	27	30	33	36
4	0	4	8	12	16	20	24	28	32	36	40	44	48
5	0	5	10	15	20	25	30	35	40	45	50	55	60
6	0	6	12	18	24	30	36	42	48	54	60	66	72
7	0	7	14	21	28	35	42	49	56	63	70	77	84
8	0	8	16	24	32	40	48	56	64	72	80	88	96
9	0	9	18	27	36	45	54	63	72	81	90	99	108
10	0	10	20	30	40	50	60	70	80	90	100	110	120
11	0	11	22	33	44	55	66	77	88	99	110	121	132
12	0	12	24	36	48	60	72	84	96	108	120	132	144

2. a. $3 \times 8 = 24$ (or $8 \times 3 = 24$)   b. $9 \times 7 = 63$ (or $7 \times 9 = 63$)

3. a. $5 \times \$9 + 5 \times \$5 = \$70$.
   b. **You need nine tables.** $9 \times 6 = 54$.
   c. They have $7 \times 4 + 4 \times 4 + 12 \times 2 =$ **68 feet in total**.
   d. **You can buy 8 shirts.** $8 \times \$6 = \$48$.

4. a. 4, 9, 6   b. 11, 6, 2   c. 7, 4, 11   d. 9, 4, 12   e. 6, 11, 2   f. 8, 2, 4   g. 6, 9, 4   h. 12, 4, 7

Chapter 4 Test

Grading

My suggestion for grading the chapter 4 test is below. The total is 22 points. Divide the student's score by the total of 22 to get a decimal number, and change that decimal to percent to get the student's percentage score.

Question	Max. points	Student score
1	8 points	
2	4 points	
3	4 points	

Question	Max. points	Student score
4	2 points	
5	2 points	
6	2 points	
Total	22 points	

1. a. 1:47; 1:57   b. 10:09; 10:19   c. 5:34; 5:44   d. 9:49; 9:59

2. a. 22 minutes   b. 39 minutes

3. a. 2 hours   b. 20 minutes   c. 43 minutes   d. 34 minutes

4. He was gone for three hours.

5. They returned on October 3.

6. It was 2:20.

Chapter 5 Test

Grading

My suggestion for grading the chapter 5 test is below. The total is 15 points. Divide the student's score by the total of 15 to get a decimal number, and change that decimal to percent to get the student's percentage score.

Question	Max. points	Student score
1	2 points	
2	3 points	

Question	Max. points	Student score
3	4 points	
4	6 points	
Total	15 points	

1. a. $8.90   b. $2.06

2. a. $1.40   b. $1.21   c. $1.16

3. a. $4.30   b. $0.70 or 70 cents.

4. a. Marsha still needs to save $16.85.
   b. The total cost is $8.51.
   c. His change is $11.49.

Chapter 6 Test

Grading

My suggestion for grading the chapter 6 test is below. The total is 31 points. Divide the student's score by the total of 31 to get a decimal number, and change that decimal to percent to get the student's percentage score.

Question	Max. points	Student score
1	4 points	
2	4 points	
3	4 points	

Question	Max. points	Student score
4	3 points	
5	8 points	
6	8 points	
Total	31 points	

1. a. 2,689   b. 4,070   c. 5,609   d. 3,902

2. a. >   b. >   c. <   d. <

3. a. 700;   8,200   b. 8,100;   8,100

4. a. 500   b. 1,400   c. 2,900

5. a. Estimate: 2,900 + 4,500 = 7,400. Exact calculation: 7,396
   b. Estimate: 7,000 − 3,000 = 4,000; Exact calculation: 4,029

6. While rounding can be done in several manners, in these problems it makes most sense to round the numbers to the nearest hundred, because that allows you to calculate the answers using mental math while keeping the estimate fairly accurate. Rounding to the nearest ten would make the mental calculations possibly too difficult, and rounding to the nearest thousand would give a very inaccurate estimate.

   a. Estimation: $2,000 − ($1,600 + $300) = $100. The exact answer: $86.

   b. Estimation: $2,600 − $700 = $1,900. The exact answer: $1916.

Chapter 7 Test

Grading

My suggestion for grading the chapter 7 test is below. The total is 38 points. Divide the student's score by the total of 38 to get a decimal number, and change that decimal to percent to get the student's percentage score.

Question	Max. points	Student score
1	7 points	
2	2 points	
3	4 points	
4	4 points	

Question	Max. points	Student score
5	5 points	
6	5 points	
7	4 points	
8	7 points	
Total	38 points	

1. A -
   B a square
   C a square
   D a rhombus
   E -
   F a rhombus
   G a rectangle

2. Area = 9 square units   Perimeter = 14 units

3. 14 cm + ? = 21 cm   or   14 cm + ? + 14 cm + ? = 42 cm
   Solution: ? = 7 cm

4. In this problem it is required that the student give the correct *unit*, not just the correct number.
   a. Perimeter = 12 m   Area = 8m^2 b. Perimeter = 28 ft   c. Area = 49ft^2 or 49 square feet

5. $2 \times 3 + 6 \times 3 = 24$ square units   OR $2 \times 3 + 3 \times 6 = 24$ square units
   OR $3 \times 2 + 6 \times 3 = 24$ square units   OR $3 \times 2 + 3 \times 6 = 24$ square units

6. Divide the shape into two rectangles (which can be done in two different ways).
   Area = 4 m × 11 m + 4 m × 8 m = 76 m^2.   OR   7 m × 4 m + 12 m × 4 m = 76 m^2.
   It is required that the student include square meters with his/her answer (m^2), not just the correct number.

7. The latter pen has a larger perimeter. Its perimeter is 320 ft, whereas the perimeter of the first pen is 280 ft. The difference is 40 ft. It is required that the student include feet with his/her answer (ft), not just the correct number.

8.   3 × (4 + 2)   =   3 × 4   +   3 × 2
      area of the   area of the   area of the
      whole rectangle   first part   second part

Chapter 8 Test

Grading

My suggestion for grading the chapter 8 test is below. The total is 20 points. Divide the student's score by the total of 20 to get a decimal number, and change that decimal to percent to get the student's percentage score.

Question	Max. points	Student score
1	2 points	
2	2 points	
3	12 points	

Question	Max. points	Student score
4	2 points	
5	2 points	
Total	20 points	

1. a. _____

   b. _____

2.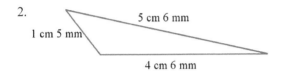
   5 cm 6 mm
   1 cm 5 mm
   4 cm 6 mm

3.

a. Mary's book weighed 350 _g_.	d. The recipe called for 2 _C_ of flour.
b. A juice box had 2 _L / qt_ of juice.	e. Mom bought 3 _kg / lb_ of bananas.
c. The airplane was flying 10,000 _ft / m_ above the ground.	f. Andy and Matt bicycled 10 _km / mi_ to the beach.
g. Erika weighs 55 _kg / lb_.	j. From Jerry's house to the neighbor's is 50 _ft / m_.
h. The shampoo bottle can hold 450 _ml_ of shampoo.	k. A cell phone weighs 4 _oz_.
i. The large tank holds 200 _gal / L_ of water.	l. A housefly measured 17 _mm_ long.

4. in   ft   yd   mi

5. a. 1 lb 11 oz   b. 3 lb 9 oz

Chapter 9 Test

Grading

My suggestion for grading the chapter 9 test is below. The total is 27 points. Divide the student's score by the total of 27 to get a decimal number, and change that decimal to percent to get the student's percentage score.

Question	Max. points	Student score
1	6 points	
2	2 points	
3	8 points	

Question	Max. points	Student score
4	3 points	
5	8 points	
Total	27 points	

1.

| a. $6 \times 7 = 42$
 $42 \div 7 = 6$
 $42 \div 6 = 7$ | b. $5 \times 11 = 55$
 $55 \div 11 = 5$
 $55 \div 5 = 11$ |

2.  or

3. a. 8, 4   b. 9, 10   c. 7, 6   d. 0, 1

4. a. 7 R4   b. 7 R3   c. 7 R1

5. a. Nine groups. $9 \times 6 = 54$.
   b. $4 \times 6 + 4 \times 10 = 24 + 40 = 64$ markers.
   c. Nine pages. $9 \times 9 = 81$, and $10 \times 9 = 90$.
   d. 27 stickers. $3 \times 9 = 27$

Chapter 10 Test

Grading

My suggestion for grading the chapter 10 test is below. The total is 27 points. Divide the student's score by the total of 27 to get a decimal number, and change that decimal to percent to get the student's percentage score.

Question	Max. points	Student score
1	4 points	
2	4 points	
3	4 points	
4	5 points	

Question	Max. points	Student score
5	3 points	
6	3 points	
7	4 points	
8	2 points	
Total	27 points	

1. a. <   b. >   c. >   d. <

2. $\frac{1}{8} < \frac{1}{5} < \frac{1}{4} < \frac{1}{2}$

3.

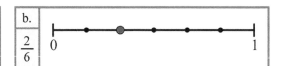

4.

5.
 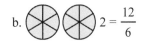 

6.
  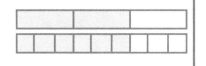

7. They eat the same amount of bread, because eating 3 pieces out of 12, and 2 pieces out of 8 signify the fractions 3/12 and 2/8, and they are equivalent fractions (both are in fact equal to 1/4).

8. It is wrong because the two wholes that we take the fractions from are not the same size. You cannot compare fractions unless the wholes they refer to are the same.

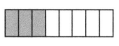

$\frac{3}{9} = \frac{3}{8}$

End-of-the-Year Test Grade 3 Answer Key

Instructions to the teacher: My suggestion for grading is below. The total is 207 points. A score of 166 points is 80%.

Grading on question 1 (the multiplication tables grid): There are 169 empty squares to fill in the table, and the completed table is worth 17 points. Count how many of the answers the student gets right, divide that by 10, and round to the nearest whole point. For example: a student gets 24 right. 24/ 10 = 2.4, which rounded becomes 2 points. Or, a student gets 85 right. 85/10 = 8.5, which rounds to 9 points.

Question	Max. points	Student score
Multiplication Tables and Basic Division Facts		
1	17 points	
2	16 points	
3	16 points	
	subtotal	/ 49
Addition and Subtraction, Including Word Problems		
4	6 points	
5	6 points	
6	4 points	
7	4 points	
8	4 points	
9	3 points	
10	3 points	
11	4 points	
	subtotal	/ 34
Multiplication and Related Concepts		
12	1 point	
13	1 point	
14	3 points	
15	3 points	
16	1 point	
17	2 points	
18	1 point	
	subtotal	/ 12
Time		
19	8 points	
20	3 points	
	subtotal	/ 11

Question	Max. points	Student score
Graphs		
21a	1 point	
21b	1 point	
21c	1 point	
21d	2 points	
	subtotal	/ 5
Money		
22a	1 point	
22b	2 points	
22c	2 points	
23	2 points	
24	3 points	
	subtotal	/ 10
Place Value and Rounding		
25	2 points	
26	5 points	
27	4 points	
28	2 points	
29	8 points	
	subtotal	/ 21
Geometry		
30	5 points	
31	2 points	
32	4 points	
33	2 points	
34	2 points	
35	3 points	
	subtotal	/ 18

Question	Max. points	Student score
Measuring		
36	2 points	
37	2 points	
38	2 points	
39	6 points	
	subtotal	/ 12
Division and Related Concepts		
40	2 points	
41	6 points	
42	3 points	
43	2 points	
44	2 points	
	subtotal	/ 15
Fractions		
45	6 points	
46	3 points	
47	2 points	
48	3 points	
49	4 points	
50	2 points	
	subtotal	/ 20
	TOTAL	/ 207

End-of-the-Year Test Grade 3 Answer Key

1.

×	0	1	2	3	4	5	6	7	8	9	10	11	12
0	0	0	0	0	0	0	0	0	0	0	0	0	0
1	0	1	2	3	4	5	6	7	8	9	10	11	12
2	0	2	4	6	8	10	12	14	16	18	20	22	24
3	0	3	6	9	12	15	18	21	24	27	30	33	36
4	0	4	8	12	16	20	24	28	32	36	40	44	48
5	0	5	10	15	20	25	30	35	40	45	50	55	60
6	0	6	12	18	24	30	36	42	48	54	60	66	72
7	0	7	14	21	28	35	42	49	56	63	70	77	84
8	0	8	16	24	32	40	48	56	64	72	80	88	96
9	0	9	18	27	36	45	54	63	72	81	90	99	108
10	0	10	20	30	40	50	60	70	80	90	100	110	120
11	0	11	22	33	44	55	66	77	88	99	110	121	132
12	0	12	24	36	48	60	72	84	96	108	120	132	144

2. a. 14, 24, 25, 36   b. 28, 40, 27, 35   c. 9, 16, 49, 32   d. 56, 30, 48, 54

3. a. 7, 5, 8, 7   b. 8, 5, 11, 7   c. 9, 7, 4, 9   d. 10, 8, 3, 3

4. a. 310, 149   b. 620, 344   c. 148, 80

5. a. 33, 5   b. 643, 45   c. 15, 378

6. a. 579. To check, add 579 + 383 = 962 using the grid.   b. 2,476. To check, add 2,476 + 4,526 = 7,002 using the grid.

7. a. 7,153   b. 792. Note the order of operations; the subtraction is done first.

8. a. △ is 294. Solve by subtracting 708 − 414.   b. △ is 824. Solve by adding 485 + 339.

9. $83

10. 160 miles. Note that the half-way point is at 150 miles. They stopped at 140 miles (10 miles before 150 miles).

11. a. 800 light bulbs   b. 736 are left. Solve by subtracting 800 − 64.

12. ⚃ ⚄

13. 5 × 25 = 125. You can solve it by adding repeatedly: 25 + 25 + 25 + 25 + 25 = 125

14. a. 48   b. 20   c. 41

15. a. 7 × 4 = 28 legs   b. 5 × 2 = 10 legs   c. 8 × 4 + 6 × 2 = 44 legs

16. 8 tables, because 8 × 4 = 32, which is more than 31. Seven tables is not enough.

17. 3 × $8 + 3 × $6 = $42

18. She needs 7 bags. (Because 7 × 4 = 28.)

19.

	a. 10:51	b. 2:34	c. 3:57	d. 5:38
10 min. later	11:01	2:44	4:07	5:48

20. a. 35 minutes   b. 5:30 AM   c. May 28

21. a. 28 hours   b. 12 hours   c. 9 hours more   d. 48 hours

22. a. $25.54   b. $9.10   c. $12.70

23. a. $2.90   b. $0.55

24. $0.60. (You can add $2.35 + $2.35 + $2.35 + $2.35 = $9.40 to find the total cost.)

25. a. 700   b. 2,000

26. a. >   b. <   c. <   d. >   e. >

27. a. 5,700; 8,600   b. 1,200; 7,800

28. a. 740   b. 990   c. 250   d. 670

29.

a. Round the numbers, then add:
 	
3,782 + 2,255
↓   ↓
3,800 + 2,300 = 6,100
	Calculate exactly: 3782
	                 + 2255
	                   6037

b. Round the numbers, then subtract:
 	
8,149 − 888
↓   ↓
8,100 − 900 = 7,200
	Calculate exactly: 8149
	                 − 888
	                   7261

30. A - rectangle   B - square   C - rhombus   D - rhombus   G - rhombus
    Also, F is a parallelogram; however that is not studied in third grade.

31. Perimeter 22 units   Area 24 square units or squares
    Note that the student should also give the "units" and "square units" or "squares", not just a plain number.

32. a. Part 1: 108 m²   Part 2: 270 m²   b. 96 m
    Note that the student should also give the units "m²" and "m" in his/her answer, not just plain numbers.

33. 9 inches.

34. a. The sides of the rectangle could be 5 and 3, or 15 and 1. Some examples below:

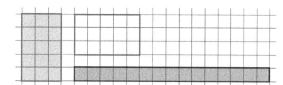

b. The sides of the rectangle could be 1 and 4, or 2 and 3.

35. $4 \times (2 + 5) = 4 \times 2 + 4 \times 5 = 28$ squares (or square units)

36. Check student's answers.

   a.

   b.

37. mm   cm   m   km

38. ounces (oz) and milliliters (ml)

39. a. feet or ft   b. cm   c. kg/lb   d. C (cups)   e. kg   f. feet or ft

40.  $3 \times 6 = 18$ $18 \div 3 = 6$

   $6 \times 3 = 18$ $18 \div 6 = 3$

41. a. 17, not possible   b. 1, not possible   c. 1, 0

42. a. 8 R1   b. 4 R4   c. 6 R5

43. Can he divide the children equally into teams of 5? **No.**
   Teams of 6? **Yes.** Teams of 7? **No.**

44. Each child paid $10.00.

45. a. $\frac{3}{8}$ b. $\frac{7}{9}$ c. $\frac{2}{4}$ d. $2\frac{2}{5}$ e. $\frac{2}{3}$ f. $\frac{9}{10}$

46. 1 = 10/10   b. 2 = 10/5   c. 4 = 24/6

47.

48.

   a. 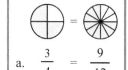 $\frac{3}{4} = \frac{9}{12}$ b. $\frac{10}{12} = \frac{5}{6}$ c. $\frac{2}{3} = \frac{4}{6}$

49. a. <   b. <   c. <   d. >

50. We cannot tell who ate more pie, because the two pies are of different sizes and it is not totally clear from the pictures which is more pie. And, even though the fraction 7/12 is more than 1/2, this thinking cannot be used here when the wholes are of different sizes.

Cumulative Reviews
Answer Keys

Cumulative Reviews Answer Key, Grade 3

Cumulative Review: Chapters 1 - 2

1. a. 574   b. 810   c. 983

2. a. XXV   b. XIX   c. LVII   d. CXLIII

3.

a. $35 - 14 - 7 + 3 = 17$ b. $35 - (14 - 7) + 3 = 31$ c. $35 - (14 - 8 + 3) = 26$	d. $(250 - 20) + (80 - 30) = 280$ e. $250 - (20 + 80 - 30) = 180$ f. $250 - 20 + (80 - 30) = 280$

4. Jill needs three more cups for her tea party.

5. a. 9, 8, 0   b. 10, 18, 32   c. 90, 80, 800   d. 0, 0, 22

6. a. $20 + 20 = 40$ b. $50 + 50 + 50 = 150$

7. a. $12 + 12 + 12 = 36$ *or* $3 \times 12 = 36$. **Tim has 36 feet of string.**
   b. **She needs six bags.** $16 + 8 = 24$; $6 \times 4 = 24$.

8.

total 400 | 250 | 150 | a. $250 + \underline{150} = 400$ $\underline{400} - 250 = \underline{150}$	total 500 | 390 | 110 | b. $\underline{390} + \underline{110} = \underline{500}$ $500 - \underline{110} = 390$

9. a. 589   b. 316   c. 258   d. 143

10. $48 + 48 + 48 = 144$. Yes, you can buy three bicycles for $150 and have $6 left over.

11.

a. rent, $256, and groceries, $387
   rent about $260
   groceries about $390
   total about $650
	b. an adult's ticket, $58, and child's ticket, $38
	   adult's ticket about $60
	   child's ticket about $40
	   total bill about $100

Cumulative Review: Chapters 1 - 3

1. Jimmy rode 44 miles each day.

2. They traveled a total of 469 miles.

3.

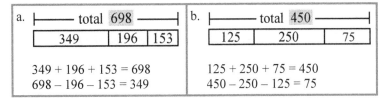

a.
$349 + 196 + 153 = 698$
$698 - 196 - 153 = 349$

b.
$125 + 250 + 75 = 450$
$450 - 250 - 125 = 75$

4. a. $4 \times 8 + 10 \times 2 = 52$. **The total cost was $52.**
   b. $12 \div 4 = 3 + 6 = 9$. **She spent $9.**

5. a. 9   b. 121   c. 67   d. 14

6. a. $2 \times 5 = 10$, $5 \times 2 = 10$ b. $2 \times 7 = 14$, $7 \times 2 = 14$

7. a. They saw 90 giraffes.   b. They saw 45 more giraffes.

Cumulative Review: Chapters 1 - 4

1.

| a. $650 + 120 = 770$ $770 - 120 = 650$ | b. $633 + 9 = 642$ $642 - 9 = 633$ |

2. a. 26   b. 56   c. 67

3. a. $13 - 9 = 4$; 16   b. $250 - 50 = 200$; 140

4. a. **Yes.** $30 - 1 - 2 = 27$; so there are 27 cupcakes left. For the afternoon tea they need $2 \times (1 + 12) = 26$ cupcakes.
   b. **9 marbles were lost.** $9 \times 12 = 108$; $108 - 99 = 9$.

| 5. a. 6:08 | b. 6:48 | c. 3:29 | d. 5:33 |
| e. 4:36 | f. 5:39 | g. 11:58 | h. 12:43 |

6.

| a. $8 \times 10 - 2 + 5 = 83$ | b. $6 + 7 \times (4 - 2) = 20$ |
| c. $3 \times 4 - 2 \times 3 = 6$ | d. $2 \times (4 + 4) \times 2 = 32$ |

7.

| a. $564 - 5 = 559$ $564 - 10 = 554$ $564 - 15 = 549$ $564 - 20 = 544$ $564 - 25 = 539$ $564 - 30 = 534$ | b. $888 + 12 = 900$ $886 + 14 = 900$ $884 + 16 = 900$ $882 + 18 = 900$ $880 + 20 = 900$ $878 + 22 = 900$ |

8. a. $3 \times 7 + 2 \times 7 = 35$. **She spent a total of 35 days.**
   b. $15 \times 5 \times 2 = 150$. **He spends 150 minutes walking to and from school.**

Cumulative Review: Chapters 1 - 5

1. a. 62   b. 61   c. 64

2. a. 32   b. 28   c. 56

3. a. 350 – 18 $\boxed{<}$ 350 – 15   b. 180 – 15 $\boxed{=}$ 190 – 25

   c. 264 + 7 $\boxed{<}$ 267 + 8   d. 62 – 27 $\boxed{>}$ 61 – 27

4. The potatoes cost $0.96 and your change is $4.04.

5. She spent 54 days in three states.

6. a. 577   b. 485

7. a. **About 215 parrot magnets.** 100 + 60 + 25 + 30 = 215.
   b. **About 205 magnets.** 40 + 85 + 55 + 25 = 205.

8. a. 35   b. 30   c. 88

9. a. 8 hours   b. 25 min.   c. 35 min.

10. a. $0.30   b. $0.91   c. $0.80

Cumulative Review: Chapters 1 - 6

1. a. 4, 8   b. 0, 9   c. 6, 9   d. 9, 9

2. a. XV   b. XXXII   c. XLVII   d. LVI

3. a. 22 till 7   b. 4 till 4   c. 12 past 2   d. 17 till 8

4. It ended at 8:15.

5. a. Estimate: 7,700 + 2,000 = 9,700   Exact: 9,760
   b. Estimate: 9,200 − 4,700 = 4,500   Exact: 4,424

6.

a. Two thousand two				b. One thousand fifteen				c. Five thousand nine hundred six			
thou-sands	hund-reds	tens	ones	thou-sands	hund-reds	tens	ones	thou-sands	hund-reds	tens	ones
2	0	0	2	1	0	1	5	5	9	0	6

7. a. $25 + ? = $69. OR $69 − $25 = ?   The unknown ? = $44. **She needs to save $44 more.**

   b. ? − $29 = $16.   The unknown ? = $45. **She had $45 before buying the gift.**

8. a. $2.68 + $4.99 + $2.95 = **10.62**
   b. Estimate: 3 × $270 = $810   OR   3 × $300 = $900   Exact: $801

Cumulative Review: Chapters 1 - 7

1. a. XVI   b. LXXXVIII   c. CXLIX   d. CCXIX

2.

a.	b.	c.	d.
$5 \times 5 = 25$	$2 \times 11 = 22$	$2 \times 7 = 14$	$5 \times 3 = 15$
$12 \times 12 = 144$	$8 \times 6 = 48$	$4 \times 12 = 48$	$1 \times 10 = 10$
$7 \times 5 = 35$	$3 \times 11 = 33$	$6 \times 7 = 42$	$8 \times 8 = 64$

3. a. The total cost was $20.50.   b. Her change was $4.50.

4.

a.	b.	c.
$8,539 \approx 8,500$	$9,687 \approx 9,700$	$5,323 \approx 5,300$
$3,551 \approx 3,600$	$1,621 \approx 1,600$	$2,399 \approx 2,400$

5. a. △ = 600   b. △ = 500   c. △ = 4,600

6.

a. 4 : 38	b. 3 : 32	c. 2 : 59
22 till 5	28 till 4	1 till 3

7. a. $14 + (1 \times 12) = 26$ c. $20 \times 4 + 8 = 88$
   b. $90 - 5 - 4 \times 4 = 69$ d. $10 \times (2 + 4) - 5 = 55$

8. She left for work at 5:55.

9. a. $3.30, $4.01   b. $0.71, $6.53   c. $15.30, $35.90

10. a. 6:45   b. 7:05   c. 12:15

Cumulative Review: Chapters 1 - 6

1. a. 4, 8   b. 0, 9   c. 6, 9   d. 9, 9

2. a. XV   b. XXXII   c. XLVII   d. LVI

3. a. 22 till 7   b. 4 till 4   c. 12 past 2   d. 17 till 8

4. It ended at 8:15.

5. a. Estimate: 7,700 + 2,000 = 9,700   Exact: 9,760
   b. Estimate: 9,200 − 4,700 = 4,500   Exact: 4,424

6.

a. Two thousand two				b. One thousand fifteen				c. Five thousand nine hundred six			
thou-sands	hund-reds	tens	ones	thou-sands	hund-reds	tens	ones	thou-sands	hund-reds	tens	ones
2	0	0	2	1	0	1	5	5	9	0	6

7. a. $25 + _?_ = $69.   OR $69 − $25 = _?_   The unknown _?_ = $44. **She needs to save $44 more.**

   b. _?_ − $29 = $16.   The unknown _?_ = $45. **She had $45 before buying the gift.**

8. a. $2.68 + $4.99 + $2.95 = **10.62**
   b. Estimate: 3 × $270 = $810 OR 3 × $300 = $900   Exact: $801

Cumulative Review: Chapters 1 - 7

1. a. XVI   b. LXXXVIII   c. CXLIX   d. CCXIX

2.

a.	b.	c.	d.
$5 \times 5 = 25$	$2 \times 11 = 22$	$2 \times 7 = 14$	$5 \times 3 = 15$
$12 \times 12 = 144$	$8 \times 6 = 48$	$4 \times 12 = 48$	$1 \times 10 = 10$
$7 \times 5 = 35$	$3 \times 11 = 33$	$6 \times 7 = 42$	$8 \times 8 = 64$

3. a. The total cost was $20.50.   b. Her change was $4.50.

4.

a.	b.	c.
$8,539 \approx 8,500$	$9,687 \approx 9,700$	$5,323 \approx 5,300$
$3,551 \approx 3,600$	$1,621 \approx 1,600$	$2,399 \approx 2,400$

5. a. △ = 600   b. △ = 500   c. △ = 4,600

6.

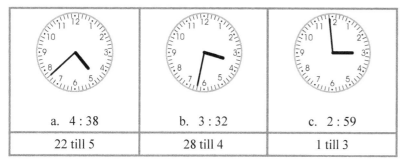

a. 4 : 38	b. 3 : 32	c. 2 : 59
22 till 5	28 till 4	1 till 3

7. a. $14 + (1 \times 12) = 26$ c. $20 \times 4 + 8 = 88$
   b. $90 - 5 - 4 \times 4 = 69$ d. $10 \times (2 + 4) - 5 = 55$

8. She left for work at 5:55.

9. a. $3.30, $4.01   b. $0.71, $6.53   c. $15.30, $35.90

10. a. 6:45   b. 7:05   c. 12:15

Cumulative Review: Chapters 1 - 8

1.

├──── total 910 ────┤	├──── total 205 ────┤
780 | 130	140 | 65
a. 780 + 130 = 910 910 − 780 = 130 910 − 130 = 780	b. 140 + 65 = 205 205 − 65 = 140 205 − 140 = 65

2. a. 4,607   b. 4,685

3. Since she started on November 3rd, she will finish on November 23rd. Notice that from November 3rd to November 23rd is 21 days, or three weeks. You will include both November 3rd and November 23rd in this count of 21 days.

4. a. <   b. <   c. >   d. >

5. a. 250   b. 720   c. 280   d. 155   e. 128   f. 29

6.

a.	b.	c.
8,509 ≈ 9,000 5,479 ≈ 5,000 7,330 ≈ 7,000	3,899 ≈ 4,000 3,809 ≈ 4,000 3,890 ≈ 4,000	5,549 ≈ 6,000 5,459 ≈ 5,000 5,594 ≈ 6,000

7. Perimeter = 14 units.   Area = 12 square units.

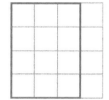

8.

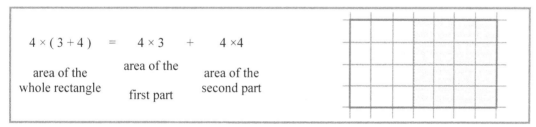

4 × (3 + 4)   =   4 × 3   +   4 × 4

area of the whole rectangle   area of the first part   area of the second part

9.

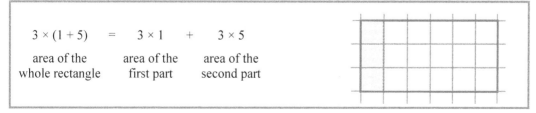

3 × (1 + 5)   =   3 × 1   +   3 × 5

area of the whole rectangle   area of the first part   area of the second part

10. a. box   b. cube   c. cylinder   d. cone

Cumulative Review: Chapters 1 - 9

1. a. **338.** Solve by subtracting 349 − 11 = 338.   b. **210.** Solve by subtracting 530 − 320 = 210.
   c. **2,100.** Solve by adding 1,600 + 500 = 2,100.

2. a. 30 ÷ 2 − 6 = ? or 6 + ? × 2 = 30. **The other side is 9 meters long.**
   b. $78 + ? = $200. **She spent $122.**

3. a. CXXIV   b. XL   c. XC   d. CCXXII

4. a. The area will be 56 ft². The perimeter will be 30 feet.

5. a. 9 × 5 = 45   b. 11 × 12 = 132   c. 9 × 9 = 81   d. 8 × 7 = 56
    6 × 5 = 30   9 × 12 = 108   7 × 9 = 63   4 × 7 = 28
    8 × 5 = 40   12 × 12 = 144   6 × 9 = 54   7 × 7 = 49

6.

| 4 × (1 + 6) | = | 4 × 1 | + | 4 × 6 |
| area of the whole rectangle | | area of the first part | | area of the second part |

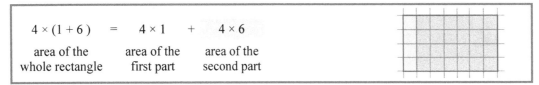

7. a. 466¢   b. $0.03   c. 205¢

8. a. 0 lb 8 oz.   b. 0 lb 12 oz.   c. 2 lb 8 oz.

9. a. She will have 11 dozen cookies, which is a total of **132 cookies**.
   b. It ends at 11:45.
   c. He traded his coins for one quarter.

Cumulative Review: Chapters 1 - 10

1.

a.	b.	c.
7 × 2 = 14	5 × 7 = 35	7 × 6 = 42
2 × 7 = 14	7 × 5 = 35	6 × 7 = 42
14 ÷ 2 = 7	35 ÷ 5 = 7	42 ÷ 7 = 6
14 ÷ 7 = 2	35 ÷ 7 = 5	42 ÷ 6 = 7

2. a. 4 × 12 = 48. **She will need 48 apples.**
   b. 36 ÷ 4 = 9. **Each piece is nine inches long.**
   c. **$1.77.** Five quarters makes $1.25. Three dimes makes $0.30. In total, you have $1.25 + $0.30 + $0.22 = $1.77.
   d. Word problems will vary. For example: Twenty-four horses are arranged into six rows for a parade. How many horses are in each row? 24 ÷ 6 = 4.

3. a. 5, 9   b. 35, 42   c. 12, 6   d. 28, 64

4. a. 7 ÷ 2 = 3, R1; 9 ÷ 2 = 4, R1   b. 13 ÷ 5 = 2, R3; 14 ÷ 5 = 2, R4   c. 21 ÷ 2 = 10, R1; 47 ÷ 6 = 7, R5

5. a. ounces (oz) or grams (g)
   b. pounds (lb) or kilograms (kg)
   c. feet (ft) or meters (m) or meters and centimeters
   d. usually feet (ft) or meters (m). Possibly miles (mi) or kilometers (km).

6. a. 5 m × 6 m = 30 m²   b. 90 m²   c. 42 m

7.

 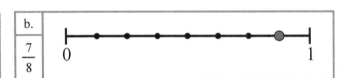

CPSIA information can be obtained
at www.ICGtesting.com
Printed in the USA
FSHW011920060719
59718FS